Climatic Changes

Climatic Changes

M. I. Budyko

American Geophysical Union
Washington, D. C.

Published under the aegis of the AGU Translation Board: Helmut E. Landsberg, Chairman; Ven Te Chow, Andrei Konradi, William H.K. Lee, John Lyman, Frank T. Manheim, David H. Miller, and Ned A. Ostenso.

Translation © 1977 by the American Geophysical Union
Russian Edition © 1974 by Gidrometeoizdat, Leningrad

ISBN 0-87590-206-5
Library of Congress Catalog Card Number: 77-81868

Printed by
Waverly Press, Inc.
Baltimore, Maryland

Foreword

At the beginning of the 1960s the opinion was voiced that noticeable, world-wide changes could take place in the climate in the very near future, due to increased industrialization and energy production. Since such changes would have a significant bearing on all fields of human endeavor, the task of utilizing this forecast in economic planning became urgent. Shortly afterwards, research on anthropogenic climatic variations commenced in several countries and the results were regularly discussed at national and international scientific meetings. These studies confirmed the pressing need for developing methods to predict future climatic variations caused by man's economic activity.

It has been recently established that it is theoretically possible to influence global climatic conditions and to prevent undesirable changes in climate by modern technical means.

A foreknowledge of changes in global climate resulting from the influence described above has become a practical necessity.

Thus it is essential to study the mechanism of climatic changes taking place at the present time, as well as those that occurred in the distant past.

Our previous book "Climate and Life" was in part devoted to this subject. Since its publication, however, much new research has been conducted emphasizing the urgency of this problem. This made it necessary to publish an additional monograph on climatic changes.

The author wishes to express his sincere gratitude to all who have read and commented on the manuscript of this work.

Leningrad, 1974.

Some small changes and additions elucidating the works carried out in 1974–76 are introduced into the American edition of the book.

Leningrad, 1976.

Figuring the nature of the times deceas'd;
The which observ'd, a man may prophesy,
With a near aim, of the main chance of things
As yet not come to life; which in their seeds,
And weak beginnings, lie intreasured.

Shakespeare, *Henry IV*

Contents

Foreword v

Chapter 1. Introduction 1
    1.1. The Purpose of This Book 1
    1.2. Sources of Information on the Climate 2
    1.3. Climates of Different Epochs 8
    1.4. Factors of Climatic Changes 17

Chapter 2. The Genesis of Climate 31
    2.1. The Energy Balance of the Earth 31
    2.2. Semiempirical Theory of the Climate 44

Chapter 3. Contemporary Climatic Changes 72
    3.1. The Warming During the First Half of the Twentieth Century 72
    3.2. Causes of Climate Warming 80

Chapter 4. Climates of the Past 98
    4.1. The Quaternary Period 98
    4.2. Pre-Quaternary Variations 111

Chapter 5. Climate and the Evolution of Living Organisms 132
    5.1. Ecological Systems 132
    5.2. Climatic Factors in Evolution 143
    5.3. Climatic Factors in Human Evolution 161

Chapter 6. Man's Influence on Climate 171
    6.1. Changes in Local Climate 171
    6.2. Changes in Global Climate 184

Chapter 7. The Climate of the Future 197
    7.1. The Outlook for Climatic Changes 197
    7.2. The Possible Ecological Crisis 219
    7.3. Climate Modification 236

Conclusion 245
Summary 246
Bibliography 248

1 Introduction

1.1 The Purpose of This Book

The subject of climatic changes has attracted the interest of many scientists. Their works have been chiefly devoted to the compilation and study of empirical data on climatic conditions during various epochs. These studies present extensive material dealing with the climates of the past.

Fewer results have been obtained, however, in studying the causes of climatic changes, although scientists working in this field have long focused their attention on these elements. Due to the lack of precision of climatic theories and the inadequacy of special observations necessary for this purpose, many obstacles have arisen that have not yet been overcome, in studying the causes of climatic variations.

Hypotheses expressed in researches on the reasons for climatic changes have very often remained unconfirmed by numerical calculations or observational data, other contentions evoking all sorts of objections. That is why there is no commonly accepted opinion today on the causes of climatic changes and fluctuations, for the current epoch as well as for the geological past.

Nonetheless, the mechanism of climatic changes is currently of much greater importance than ever before. Several studies have found that man's economic activity has begun to affect climatic conditions on a global scale and that this influence is increasing very rapidly.

This makes it highly essential to find ways and means of predicting climatic variations that would so deteriorate environmental conditions as to make them hazardous for mankind.

It is obvious that such predictions cannot be based exclusively on empirical data about past climatic changes, though such data could be useful for extrapolating currently observed climatic variations for a projection of future conditions. This method of forecasting, however, is suitable only for very limited time intervals, since the numerous factors affecting the climate are highly variable.

In as much as the human economy is increasingly influencing atmospheric processes, a physical theory of climatic changes must be used in order to develop a reliable method of predicting the climate of the future.

The achievements of modern climatology research are great, but it must be borne in mind that the quantitative models of the meterological regime are highly approximate, and the principles they are based on involve various important limitations.

Obviously all available empirical data on this subject are of particular importance in developing and verifying approximate theories of climatic variations.

This likewise applies to the study of probable consequences from changes in the global climate that may be expected in the foreseeable future.

Prospects for shedding light on the mechanism of climatic changes are enhanced by the rapid development in recent years of a new branch of science called physical climatology, which applies the methods of theoretical and experimental meteorology for studying climatic conditions. Many research reports in this field explain various regularities in climate genesis, i.e. the conditions of climate formation in various regions of the globe.

It may be assumed that the use of modern methods in physical climatology will facilitate the study of the causes of climatic variations on a wider basis than the majority of research work carried out in this field to date.

This book is devoted to these problems and is based mainly on data acquired by the author and his colleagues at the A. I. Voeikov Main Geophysical Observatory.

Chapter 1 gives a brief description of climatic conditions in the past and a survey of various opinions on the causes of climatic changes. Chapter 2 presents a description of a semiempirical theory of climate, which is applied in Chapters 3 and 4 to clarify the causes of natural climatic changes of the past and present. In Chapter 5 we discuss the influence of climatic changes on the evolution of living organisms.

The final chapters of this book are devoted to the effects of man's economic activity on the climate at the present time and in the future. These chapters deal with the possibility of an ecological crisis arising as a result of the anthropogeneous variations of climatic conditions as well as with ways and means of preventing climatic variations that could prove hazardous to man's economy.

1.2 Sources of Information on the Climate

We glean the greatest amount of knowledge about climatic conditions on our planet from what has been called the era of instrumental observations, during which systematic measurements of a number of meteorological variables were made by means of standard instruments at a large number of stations. It is difficult to name the exact beginning of this era because it took a long time to set up and develop the vast network of meteorological stations which now exist throughout the world.

Although the prototypes of most meteorological instruments were developed during the seventeenth and eighteenth centuries, the number of meteorological stations conducting continuous observations remained very small up to the middle of the nineteenth century, especially in tropical countries and sparsely populated regions at other latitudes. By the latter half of the nineteenth century, a more or less numerous network of such stations was already functioning on all continents except the Antarctic. Now there are over 10,000 such stations and their programs include measurements of temperature, humidity, precipitation, pressure, wind velocity, and several other meteorological elements. There also exists a system of actinometric observations which is of vast importance for studying the climate and its variations.

Today the actinometric network comprises about 1,000 stations that conduct observations of incident solar radiation and its transformations at the earth's surface.

In recent years, meteorological satellites have facilitated the study of the climate by supplying very important material, including data on clouds and the radiational regime at the outer boundary of the atmosphere.

Mention must be made of special research being carried out by a limited number of stations, such as observations of the composition of the atmosphere and of the solid and liquid particles it contains, atmospheric aerosol. These observations commenced only a few decades ago and their importance for the study of climatic changes cannot be overestimated.

Yet it must be noted that the existing system of meteorological observations throughout the world is, in many respects, inadequate for studying climate and its variations. For instance, there is comparatively little information about atmospheric conditions over the vast part of the earth's surface covered by oceans. There are at present relatively few weather ships (and those situated chiefly in the North Atlantic) that conduct observations of all important meteorological elements in certain regions. Data on the meteorological regime of other regions of the world's oceans are supplied mainly by passing ships. This information is inferior to that obtained by meteorological stations on the continents, both in quantity and quality.

Many meteorological parameters vital for climatic studies, such as turbulent heat flow in the atmosphere, evaporation, heat consumption during evaporation, etc., are not recorded at the majority of meteorological stations, although methods for doing so have been developed and are being applied by a small number of research stations.

No direct methods for observing horizontal heat advection, heat release during condensation of water vapor, and other climatic elements have as yet been developed.

Despite its limitations, the world's system of meteorological observations has yielded a great deal of material in the past century enabling scientists to study the variations of the atmosphere's thermal regime near the earth's surface, precipitation, pressure, winds, and many other meteorological elements.

Earlier data for the meteorological study of certain regions are also available, but the territory covered by these observations, carried out during the first half of the nineteenth and particularly during the eighteenth century, comprises only a small part of the earth's surface.

A certain amount of information on climatic conditions prevailing in the period of two to three thousand years before the era of instrumental observations may be found in the writings of contemporaries. These findings vary widely in character; some were obtained in the course of several decades and involve a detailed description of day-to-day weather changes, while some present brief notes of great catastrophes that attracted universal attention, such as droughts, hurricanes, torrential rains, etc.

Since this sort of information refers only to a very limited number of regions, it may be applied mainly to the study of climatic changes in those particular locations.

Considerably more important for the study of the climates of various epochs are data on the natural conditions of the past. Since such processes as the formation of sedimentary deposits, the weathering of rocks, the development of reservoirs and the existence of living organisms depended on atmospheric factors, available data on these processes allow us to estimate climatic conditions during given periods.

It is very difficult indeed to interpret data on the natural conditions of remote epochs to clarify the climatic regime, though some of these data are of fundamental importance. One such difficulty arises due to the need for applying the principle of uniformitarianism to these studies. This means accepting the assumption that in the past the relations between the climate and other natural phenomena were similar to those existing today. While this approach is by no means beyond contention, the great variety of natural phenomena depending on the climate allows us to independently test this hypothesis, by reconstructing the climatic conditions of the past in accordance with divers paleogeographic indications. There is no doubt that the general laws and patterns of climatic conditions during the geological past, as established in paleogeographic research, hold true, although certain results of these investigations are in dispute and require further confirmation.

An essential complement to the paleogeographic data in studying climatic conditions of the past are materials on the paleotemperatures obtained as a result of analyzing the isotopic composition of organic remains. The task of estimating the precision of these data and interpreting them correctly is fraught with great difficulties, but these are gradually being overcome with the development of methods for paleotemperature research.

The study of past climates involves the application of information on the structure of sedimentary rocks, and on geomorphological properties, as well as data on fossil flora and fauna.

The former approach is based on the currently well-established relations between the lithogenetic and the climatic factors. For example, under the conditions of a hot and humid climate, an intensive chemical weathering of rocks takes place leading to the decay of chemically nonresistant minerals. Chemical weathering is less intensive in a hot and dry climate but the destruction of rocks by winds and temperature fluctuations begins to play a serious part. In a cold climate, chemical weathering is even less marked and physical weathering predominates, so that the chemically nonresistant minerals are preserved.

Humidity has a telling influence on the amount and structure of sedimentary deposits. In dry regions the amount of such deposits is scarce, while in humid regions it is much greater, and includes alluvial deposits to a great extent.

The process of coal formation is closely connected with climatic conditions, which means that data on fossil coal can be used for reconstructing past climates. The interpretation of these data, however, presents certain difficulties because the dependence of coal accumulations on climate during different periods of time varied greatly in accordance with the changing character of vegetation from which it was formed. For instance, while many of the coal beds of the Devonian period were formed under dry climatic conditions, coal accumulation during the Permian and Carboniferous times was associated with comparatively humid climatic conditions.

When studying the climates of the past we also have at our disposal data on sediments of limestones, dolomites and salt-containing deposits which are valuable aids in reconstructing climatic conditions of ancient reservoirs.

In some cases, data on the structure of sedimentary rocks is used for estimating seasonal climatic conditions. The best known example of this is the study of clays formed in regions near the periphery of continental glaciations. During the summer melting of glaciers, the streams washed from their surfaces a large amount of coarse, fragmented material, and during the cold season, a smaller amount of fine-grained, argillaceous matter was deposited in these regions. The stratified structure of varved clays enables us to estimate the duration of their formation period.

In order to study the regime of atmospheric precipitation and the formation of the ice cover, we have wide recourse to geomorphological data, including information on the position of the ocean coastline. These allow us to estimate the water loss and gain by the ocean due to the formation and melting of continental galciation.

A major source of information on the development of glaciers are data on changes in the surface relief under the influence of glaciation. Another important aid in determining past climates are data on the altitude of the snow-line in the mountains, as its position depends on the temperature and precipitation.

Pertinent information on humidity conditions may be gleaned from data on ancient lakes and river valleys. For instance, traces of numerous lakes and rivers on the territory of present-day deserts attest to the drastic humidity changes that occurred after the epoch in question.

Data on the level oscillations of land-locked reservoirs, the Caspian Sea, for example, allow us to estimate the water inflow into this reservoir and, consequently, the amount of precipitation in the basin of its feeding rivers during various periods of time.

For the study of humidity and thermal regimes of the past, information on the nature of fossil soils may be used. Traces of permafrost in the soil is of importance for reconstructing zones with a cold climate.

Likewise important for paleoclimatic investigations are data on the nature of the erosion process, which also depends greatly on climatic conditions, and particularly on humidity.

Data on the geographical distribution of living organisms, and especially of vegetation, which is highly dependent on climatic conditions, are also a major contributing factor in the study of past climates. The application of this method yields more reliable results for the not-too-distant past when the vegetation differed only slightly from its modern forms, and climatic conditions affected its distribution in the same way as today.

For reconstructing the climates of more remote epochs, it is difficult to apply the relation between the vital activity of present-day vegetation and meteorological factors, and this limits the use of evidence on the distribution of plants in paleoclimatic studies.

One great achievement of the last decades was the use of a method for analyzing fossil pollen and spores, from which we get a pretty good idea of the vegetation cover of a given region. This method is also easier to apply to epochs in which the vegetation was similar to that of today.

Annual growth rings on trees tell us a great deal about the climatic history of the past. These rings reveal short-term climatic fluctuations while their structure is an indicator of general climatic conditions. For example, weakly developed growth rings on trees in carbonaceous swamps indicate the absence in this case of distinct seasonal variations.

It is more difficult to utilize material on fossil fauna since the dependence of the geographical distribution of animals on the climate is, as a rule, not as pronounced as that of vegetation. Nevertheless, these data, and particularly data on poikilothermal (or cold-blooded) animals, are a valuable complement to other methods for the study of climatic variations. The interpretation of data on the distribution of animals as well as of vegetation for the purpose of estimating climatic conditions is based on the assumption that the influence of climate upon the life of the past organisms was similar to the influence upon the related forms existing today.

In paleoclimatology it is of special importance to apply methods for determining paleotemperatures from the content of the oxygen isotope O^{18} in the fossil remains of aquatic animals. It has been established that the ratio of O^{18} to O^{16} in shells of mollusks and other remnants of sea organisms depends on the temperature under which these organisms lived.

The determination of paleotemperatures from the isotopic content of specimens under study made it necessary to develop highly sensitive mass spectrometers and to solve a number of other technical problems. Beginning with the middle of our century, these methods of direct measurements of paleotemperature gained wide recognition in the study of recent climatic conditions and those that existed over hundreds of millions of years ago.

When evaluating the results of studying climatic changes by means of all the above described methods, it should be mentioned that, with the exception of a relatively short recent period, nearly all the data on past climates refers to the air temperature at the earth's surface, to that of land and bodies of water, and also to the moisture conditions on the continents.

Several other climatic elements may also be gleaned from paleographic data: in some cases we can determine the direction of prevailing winds from the forms of fossil dunes and barkhans; but this information is comparatively negligible.

Data on the air temperature near the earth's surface is of paramount importance in studying climatic conditions also in the era of instrumental observations as well, for two reasons: firstly, because of the close relation between the air temperature and the components of the heat budget which greatly affect the climate, and secondly, because nearly all other elements are registered at meteorological stations with a relatively lower precision than the temperature, thus limiting the use of these data in the empirical study of climatic variations.

1.3 Climates of Various Epochs

Modern Climate. For comparison with climatic conditions of the past, let us briefly describe modern climatic conditions, with the emphasis on two meteorological elements — air temperature at the earth's surface and the amount of precipitation.

Table 1 represents data on mean latitudinal temperatures at sea level for January and July and on atmospheric precipitation over the entire year. We see from this table that the mean temperature varies by approximately 70°C from the maximum value at the equator to the minimum value at the south pole. The reason for the temperature variations is discussed in detail in the next chapter; here we only mention that due to the spherical form of the earth, the solar radiation incident upon the outer boundary of the atmosphere varies greatly with latitude.

At high latitudes where the air temperature is below the freezing point for water during the whole or greater part of the year, there is a permanent ice cover. In the Arctic the main part of this cover consists of

Table 1
Air Temperature and Precipitation

Latitude (°N)	Temperature (°C)		Precipitation (cm/yr)	Latitude (°S)	Temperature (°C)		Precipitation (cm/yr)
	January	July			January	July	
90–80	−31	−1	19	0–10	27	25	147
80–70	−25	2	26	10–20	26	24	129
70–60	−22	12	97	20–30	25	18	85
60–50	−10	14	72	30–40	20	14	92
50–40	−1	20	78	40–50	12	8	102
40–30	11	26	77	50–60	5	1	97
30–20	19	28	70	60–70	0	−12	67
20–10	25	28	117	70–80	−8	−30	25
10–0	27	27	192	80–90	−13	−42	11

comparatively thin sea ice, while in the Antarctic it consists of a thick continental glaciation covering almost the entire surface area of the continent.

Along with the strong meridional variations, the mean temperature at the earth's surface in the majority of latitude zones varies significantly with longitude. These changes are caused mainly by the distribution of continents and oceans.

In middle and high latitudes the air temperature in the summer is considerably lower over the ocean than over the continents, while in the winter it is higher. This is due to the high thermal capacity and conductivity of water, causing the oceans to absorb a great amount of solar energy in the summer and to discharge it during the winter season. A telling effect on the thermal regime is produced by ocean currents, particularly by warm surface currents that transfer a large amount of heat during the cold season from low and middle to high latitudes, where it is released and contributes to the maintenance of comparatively high air temperatures.

Oceans have a strong climatic influence on continents in middle and high latitudes. Here the so-called maritime climate prevails, characterized by a small temperature range throughout the year. In the nontropical parts of continents, where the influence of the ocean is less noticeable, seasonal temperature variations increase sharply, corresponding to the conditions of a continental climate.

As to the distribution of the zonal mean precipitation, we find a primary maximum in the equatorial region, a decrease in subtropical latitudes, secondary maximums in middle latitudes, and a decrease with increasing latitude to the polar region.

The variations of the zonal mean precipitation are governed by the distribution of the average air temperature and the peculiarities in the atmospheric circulation.

Other conditions (including relative air humidity being equal, the total precipitation increases with temperature since the amount of water vapor consumed in the condensation process also increases. It is obvious that if no other factors had any serious effect on precipitation, its latitudinal distribution would have a single maximum in the lower latitudes.

However, for falling precipitation, the field of vertical air motions is of paramount importance because it affects the water vapor transfer across the condensation level, leading to the formation of clouds and precipitation.

General atmospheric circulation is closely related to the geographical distribution of semi-permanent pressure systems, among which the most important are the low pressure band near the equator, high pressure regions in the tropical latitudes outside of the equatorial bands and in subtropical latitudes, and regions with a high cyclonic activity in middle latitudes. Since descending air currents prevail in the high pressure zones, the amount or precipitation in this region decreases, producing two minimums on the curve of the latitudinal distribution of precipitation.

The considerable intensity of ascending air currents in the equatorial and parts of middle latitudes is responsible for increased precipitation in these regions.

The earth's largest deserts lie in the subtropical high pressure zone, where the amount of precipitation is very low. The precipitation deep in the continents in middle latitudes also decreases as the wind currents carry very little water vapor from the ocean that far, thus leading to a decrease in the relative air humidity and a weakening in the condensation processes of water vapor.

Hence, the zones of humid climate on the continents are situated mainly in the equatorial latitudes and in regions of oceanic climate in the middle and high latitudes. Conditions of insufficient humidity prevail in high tropical and subtropical latitudes and in regions of continental climate.

Time Scale. In discussing available empirical data on climatic fluctuations, we shall deal with four time intervals whose spans increase as we go down into the ages. The first of these intervals refers to the period of instrumental observations a little more than one century old. The second interval, the Holocene, covers an interval following the end of the last glaciation, that is, a little more than 10,000 years. The third interval, the Pleistocene, lasted about 2 million years. When studying the Pre-Quaternary, or fourth interval, we shall concentrate mainly on the last 200 million years, for which data on climatic conditions are more reliable.

Note that the duration of each interval is 100–200 times greater than the previous one. This division into intervals is dictated by a decrease in the amount and accuracy of data on climatic conditions for more remote epochs in comparison with less remote.

When discussing past climatic conditions, we should take into account information on the sequence and duration of various stages in the geological history of the earth. Some data on these stages are given in Table 2.

The Tertiary period is usually divided into two parts: the Paleogene (which includes the Paleocene, Eocene and Oligocene), and the Neogene (which includes the Miocene and Pliocene). Table 2 does not give the names of the epochs in the Pre-Cenozoic time, and these are not referred to in the text.

The Paleozoic era was preceded by the Proterozoic, which lasted at least two billion years. The Post-Proterozoic time is often called the Thanerozoic.

The geochronological scale presented in Table 2 is based on successive changes in lithogenesis, flora and fauna established during the study of the earth's crust. The division of the earth's geological history into periods which takes these changes into account is inevitably fraught with a margin of uncertainty; therefore, discrepancies as to the most reasonable construction of the geochronological scale can hardly be avoided. For instance, a consideration of the Holocene as a separate epoch of the Quaternary period, is often subject to dispute, because its duration comes to less than 1% of the total span of the Quaternary.

Without pausing to discuss these discrepancies, which are of no importance in the study of past climates, we would like to point out the limited precision of the data given in Table 2 on the duration of various

Table 2
Geochronological Scale

Period	Epoch	Duration of period (million years)	Absolute age of beginning of period (million years)
	Cenozoic era		
Quaternary	Holocene Pleistocene	2	2
Tertiary	Pliocene Miocene Oligocene Eocene Paleocene	64	66
	Mesozoic era		
Cretaceous		66	132
Jurassic		53	185
Triassic		50	235
	Paleozoic era		
Permian		45	280
Carboniferous		65	345
Devonian		55	400
Silurian		35	435
Ordovician		55	490
Cambrian		80	570

geological periods and, consequently, on their absolute age. The probable error of this dating, in any event, is not less than several per cent of the values given here.

Contemporary Climate Changes. For a brief description of climatic conditions during various epochs we shall resort to material contained in a series of monographs published in recent years (Bowen, 1966; Flint, 1957; Inadvertent Climate Modification, 1971; Lamb, 1973; Mazkov, 1960; Rubinshtein and Polozova, 1966; Schwarzbach, 1961; Sinitsyn, 1967; Velichko, 1973).

The most significant climatic change in the period of instrumental observations commenced at the end of the nineteenth century. It was marked by a gradual increase in the air temperature over all latitudes in the northern hemisphere, during all seasons of the year, while the greatest warming occurred in high latitudes during the cold half of the year.

The warming was continuous throughout the 1910s and 1920s and reached its maximum in the 1930s when the mean air temperature in the northern hemisphere registered in increase of about 0.6° as compared to the end of the nineteenth century. In the 1940s this warming gave way to a cooling process which has continued to recent time and resulted in lowering the mean air temperature by 0.3–0.4°.

Although data on modern climatic changes in the southern hemisphere are not as clear as those for the northern hemisphere, there are grounds to believe that, there too, a warming occurred in the first half of the twentieth century.

In the northern hemisphere the rising air temperature was accompanied by a decrease in the surface area of the polar ice cover, a recession of the permafrost boundary into higher latitudes, a northward expansion of the forests and tundra boundaries, and by other changes in natural conditions.

Of great importance was the marked change in the regime of atmospheric precipitation during the warming. The amount of precipitation in several regions of insufficient humidity dropped even lower during the warming, particularly in the cold period of the year. This resulted in a decrease of river runoff and a drop in the water level of several land-locked reservoirs.

In the 1930s the drop of the water level of the Caspian Sea, caused mainly by the reduced discharge of the Volga River, was particularly noticeable. At the same time the warming brought more frequent droughts in vast territories of intracontinental regions at middle latitudes in Europe, Asia and North America. This climatic change affected the economies of many countries.

The Holocene. The Post-Quaternary climatic fluctuations comprise several large climatic fluctuations.

The Würm glaciation reached its maximum development about 20,000 years before our time. In several thousand years the area of this glaciation was considerably reduced. The succeeding epoch was marked by a comparatively cold and humid climate in the middle and high latitudes of the northern hemisphere. About 12,000 years before our time, there was a significant warming (the Alleröd epoch), which, however, soon changed to cooling. During these climatic fluctuations, the summer air temperature in Europe underwent a change of several degrees.

Later, the warming resumed and the last big glaciations in Europe and North America disappeared 5,000–7,000 years before the present. During this period the post-glacial warming reached its peak. It is assumed that about 5,000–6,000 years ago the air temperature in middle latitudes in the northern hemisphere was about 1–3° higher than it is today.

Changes in the atmospheric circulation apparently occurred simultaneously with the northward shift of the polar-ice boundary. The subtropical high pressure zone also moved to higher latitudes, resulting in the expansion of drought areas in Europe, Asia and North America. At the same time precipitation increased in low latitudes, over the region which are desert today. During this epoch the climate of the Sahara was comparatively humid and it possessed rich flora and fauna. Later there was a tendency toward cooling which was particularly noticeable in the early half of the first millenium of our epoch. Parallel with the change in the thermal regime, the precipitation regime also varied, gradually reaching its present condition.

A noticeable warming took place at the end of the first and the beginning of the second millenium A.D. During this time the polar ice retreated into high latitudes, making it possible for the Vikings to colonize Greenland and leading to their discovery of the continent of North America. A later cooling led to a fresh encroachment of the ice cover, and the Greenland colony, no longer able to communicate with Europe, perished.

The cooling which commenced in the thirteenth century and reached its culmination at the beginning of the seventeenth century was accompanied by the growth of mountain glaciers, wherefore it is sometimes referred to as the Little Ice Age. Later another warming and a recession of glaciers took place.

In the eighteenth and nineteenth centuries climatic conditions were more or less the same as they are today.

The Pleistocene. During the Pleistocene the climate differed greatly from the previous conditions of the Mesozoic era and the Tertiary period, when the thermal zones were barely distinguishable. In the course of the Pleistocene the cooling in middle and high latitudes intensified, bringing about the formation of large continental glaciations. We know very little about the number of these glaciations or their ages; it is generally assumed that the first of them originated about 700,000 years ago.

Research work carried out in the Alps in the last century revealed evidence of four major European glaciations: Guntz, Mindel, Riss and Würm. It became known later that each of these glaciations had several stages with intervals of recession. The Alpine glaciations coincided with the development of the ice cover on the plains of Eurasia and North America, stretching over vast areas in high and middle latitudes.

It would seem that the periods of glacier advance and recession comprised the lesser parts of the total duration of the Pleistocene, while the comparatively warm interglacial epochs were of longer duration. In the course of the warm epochs the continental ice cover remained only in mountain regions and high latitudes.

It has been established that the advance and recession of glaciers in Europe, Asia and North America were more or less synchronous, and this also applies to the ice ages in the northern and southern hemispheres.

The continental ice cover expanded during the ice ages mostly in the regions with a humid oceanic climate, while in the comparatively dry climate of northern Asia it occupied a very small area. During the most severe glaciations the continental ice cover in the northern hemisphere reached an average latitude of 57°N, while in certain places it extended as far as 40°N. Its thickness over a considerable part of the area came to several hundred meters, even reaching a kilometer or more in some places.

There is no doubt that during the expansion of continental glaciation the boundary of the polar sea ice also migrated to lower latitudes. This considerably increased the total area of the permanent ice cover on the planet.

During each advancement of glaciers the snow-line in the mountains which were beyond the glaciated areas dropped several hundred meters, and sometimes a kilometer or more.

In addition, the area of permafrost expanded significantly during the ice ages. The permafrost boundary shifted several thousand kilometers into the lower latitudes.

A characteristic factor of the ice ages was the lowering of world sea level by 100–150 meters as compared to its present one; this was the result of enormous quantities of water being "locked up" in the thick continental glaciations. In the warm interglacial epochs sea level rose several dozen meters as compared with the present level. The fluctuations of sea level changed the surface area of the oceans by several per cent.

The climate in the ice ages was marked by a noticeable drop in air temperature over the entire globe. This temperature drop averaged several degrees and increased with latitude. During the interglacial periods the temperature was higher than today, though this warming was not as marked as the cooling.

The picture of precipitation changes associated with glaciations is not so clear. Available data indicate that glaciations exercised a varying influence on humidity in different parts of the globe. This testifies to changes in the atmospheric circulation system due to glaciation and is related to fluctuations in the temperature gradients between the poles and the equator.

At the height of the glacial development there was a decrease in precipitation over many continental regions, apparently due to the reduced evaporation from the ocean surface, an extensive part of which was covered by ice.

It should be noted that the first continental glaciations at high latitudes presumably took place as far back as the Miocene epoch. Their centers were situated in mountainous regions.

The development of large continental glaciations not connected with mountain systems is a characteristic feature of the Pleistocene period.

Pre-Quaternary Time. Our knowledge of the Paleozoic climatic conditions is extremely vague. Apparently, during the major part of this era (which existed 570-235 million years before present), the climate throughout the globe was very warm and humidity conditions on the continents extremely diverse. By the end of the Paleozoic era and around the Carboniferous and Permian periods, a glaciation developed that spread over a vast area, the major part of which is not tropical. It is difficult to estimate the geographical position of this glaciation at that time because the continents and the earth's poles have since shifted greatly. Climatic conditions in other regions of the globe were typically quite warm during this Permo-Carboniferous glaciation. The thermal zones in the Permian period became pronounced and the regions of dry continental climate vastly expanded.

The Mesozoic climate (235–66 million years before present) was rather uniform. Most of the earth resembled the tropics today; while in high latitudes the climate was more moderate, although very humid, and fluctuated only slightly with the seasons. The humidity conditions on the continents during the Mesozoic era were apparently more uniform as compared to recent times, though there were zones of insufficient and excess moisture.

At the end of the Cretaceous period the hot zones diminished while the dry regions expanded. During the transition to the Cenozoic period there were no marked climatic changes. The second half of the Tertiary period (from middle of Oligocene) was characterized by the beginning of a temperature drop that was more pronounced in the middle and especially in high latitudes. From that time on a new climatic zone emerged in the higher latitudes and gradually expanded, its meteorological regime resembling the climatic conditions in the middle latitudes today. In this zone the air temperature dropped below 0°C in winter, resulting in the formation of a seasonal snow cover. At the same time, the continental climate became more extreme in the central parts of the continents.

The cooling did not proceed at an even rate; at certain periods there were warm spells which, however, did not affect the general tendency toward the formation of thermal zones caused by the temperature drop in high latitudes. This process accelerated during the Pliocene when the first large continental glaciation, still existing today, emerged in Antarctica.

The climate of the Pliocene was warmer than ours but came closer to it than that of the Mesozoic and the early half of the Tertiary period.

A study of successive climatic variations over the periods for which more or less reliable data are available, leads us to the conclusion that regimes with strongly pronounced thermal zones are not typical of climatic conditions on our planet.

The great temperature gap between the poles and the equator, which has existed since the end of the Tertiary period and increased particularly during the ice ages, is typical for only the small fraction of time that has elapsed since the onset of the Paleozoic era. During the last 600 million years there was only one large Permian-Carboniferous glaciation (besides the Quaternary glaciations), and its duration was also small as compared to the time that elapsed since the beginning of the Paleozoic era.

At the same time we cannot be certain that the big difference in temperature between the poles and the equator existed when this glaciation developed. (This question is dealt with in more detail in Chapter 4.)

Data on climatic conditions in the Pre-Cambrian time are not very reliable. For instance, there is no consensus of opinion on the interpretation of the till-like rock which some scientists consider to be deposits of large glaciations. Though we may assume that glaciations occurred during the Proterozoic era, their position relative to the earth's pole at that time is totally unclear.

Thus, the climatic conditions of the last comparatively brief interval in the earth's geological history with distinct thermal zones probably have no close analogy with any past climates. They differ drastically from the conditions that prevailed in the course of the greater part of the last 600 million years, when hot or warm climate prevailed at all latitudes, when there were small temperature gradients between the poles and the equator, and a small variation of the annual temperature beyond the tropical zone.

1.4 Factors of Climatic Changes

The reasons for climatic changes may be divided into two groups. The first group covers external factors that undoubtedly influence the climate, although opinions differ as to the scale and character of this influence. The second group covers the factors whose influence on the climate is not beyond dispute; some authors acknowledge and others deny the role of these factors in climatic changes.

The factors of the first group include changes in atmospheric composition, in the form of the earth's surface, and in astonomical elements defining the position of the earth's surface relative to the sun.

Changes in Atmospheric Composition. Contemporary investigators devote special attention to the climatic implications of two variable components of the atmosphere, i.e. the concentration of carbon dioxide, and the amount of aerosol ejected into the atmosphere during volcanic eruptions.

Over a century ago Tyndall (1861) pointed out that since carbon dioxide and water vapor absorb long-wave radiation in the atmosphere, the changes in CO_2 concentration may lead to climatic fluctuations.

Later this problem was studied by Arrhenius (1896, 1903) and Chamberlain (1897, 1898, 1899), who assumed that the change in the amount of atmospheric carbon dioxide could have caused the Quaternary glaciations.

Arrhenius studied the absorption of atmospheric radiation and suggested a quantitative model for determining the temperature at the earth's surface as a function of atmospheric properties. By applying this model, Arrhenius deduced that a 2.5 to 3-fold increase of carbon dioxide raises the air temperature by 8–9°C, while a decrease of 38–45% lowers the temperature by 4–5°C.

On the basis of geological data Arrhenius found that the carbon dioxide left in the atmosphere today is only a small fraction of the amount consumed by carbonaceous sediments during their formation in the past. Furthermore, Arrhenius came to the conclusion that the concentration of carbon dioxide in the atmosphere could have changed to a large extent and that these changes greatly affected the air temperature thus facilitating the emergence and disappearance of glaciations.

Chamberlain's works were mainly devoted to the geological aspects of this problem. He noted from a study of the balance of carbon dioxide in the atmosphere that the inflow of carbon dioxide from the lithosphere changed significantly with the level of volcanic activity and other factors.

The consumption of carbon dioxide during geological processes was also modified to a large extent, depending, in particular, on the surface area of rocks subject to atmospheric erosion. With the increase of the surface area the consumption of carbon dioxide caused by the weathering process increased. Chamberlain assumed that the glaciations originated as a result of the intensive process of mountain formation and the rising of continents which led to an increase of erosion and of the weathered rock surface, and, consequently, to a decrease in the concentration of carbon dioxide in the atmosphere. To support his hypothesis, Chamberlain carried out calculations; these, however, cannot be considered a complete quantitative model of the given process.

Later the carbon dioxide hypothesis of the origin of Quaternary glaciations became doubtful because it was discovered that in the 13–17 μm band there is a considerable absorption of radiation by water vapor and not only by carbon dioxide, thus reducing the influence of carbon dioxide amount on the thermal regime.

By taking this effect into consideration, Callender (1938) obtained smaller values of temperature variations at the earth's surface with fluctuations of the carbon dioxide concentration, than did Arrhenius. According to Callender, a doubling of the carbon dioxide increases the air temperature by 2°C, and this dependence weakens with the growth of carbon dioxide concentration.

Callender suggested that the warming of the climate in the first half of the twentieth century was connected with an increase in the carbon dixoide concentration in the atmosphere caused by man's industrial activities.

Callender proved the importance of considering the absorption of a long-wave radiation by water vapor, for correctly evaluating the effect of carbon dioxide concentrations on the temperature. However, Plass (1956) and Kaplan (1960) did not take this into account when carrying out similar calculations.

Plass has obtained comparatively large temperature variations at the earth's surface: the temperature increases by 3.6° when the carbon dioxide concentration doubles, and decreases by 3.8° when the concentration drops to half of its present amount. This has enabled him to revert to the carbon dioxide hypothesis on the origin of Quaternary glaciations. Plass suggested that the changes in the carbon dioxide concentrations in the atmosphere and ocean were of an auto-oscillatory nature.

Plass claimed that when there is a small initial decrease in atmospheric carbon dioxide, the temperature at the earth's surface drops, resulting in the growth of the polar ice mass. This causes a reduction in the volume of ocean water, which leads to an increase in the atmospheric carbon dioxide, and consequently, to a new warming. This, in turn, causes the ice to melt, the volume of water in the ocean to increase, and the amount of carbon dioxide in the atmosphere to drop, whereupon a new cooling commences.

Kaplan showed that the cloud cover considerably weakens temperature variations at the earth's surface caused by carbon dioxide fluctuations, as compared to the estimates given by Plass.

Kondrat'ev and Niilisk (1960), as well as Möller (1963) stressed the necessity of taking into account the absorption of long-wave radiation by water vapor in calculating the atmospheric thermal regime. The values of temperature variations obtained by Kondrat'ev and Niilisk were much smaller than those obtained by Plass and Kaplan.

Möller pointed out that the variations in air temperature are generally accompanied by a change in absolute humidity, while the relative humidity remains more or less constant. The increase of the absolute air humidity with temperature enhances the absorption of long-wave radiation in the atmosphere, and this leads to further temperature rises. Möller has established that the temperature increase at the earth's surface comprises 1.5° when the concentration of carbon dioxide doubles while the absolute air humidity remains constant, whereas with constant relative humidity this increase is several times greater. At the same time Möller noted that the influence of carbon dioxide on the thermal regime can be compensated by comparatively small changes in the absolute air humidity or cloudiness.

A detailed investigation of the effect of carbon dixoide on the air temperature was carried out by Manabe and Wetherald (1967). They pointed out the inaccuracy of the calculations made by Möller, who estimated the thermal regime variations in the atmosphere only on the basis of the thermal balance of the earth's surface, without taking into account the thermal balance of the atmosphere as a whole.

Manabe and Wetherald calculated atmospheric temperature profiles, taking into consideration the absorption of long-wave radiation by water vapor, carbon dioxide and ozone. In their work they relied on the profiles of relative air humidity based on empirical data. It was assumed that the temperature distribution was determined by conditions of local radiation balance, if the vertical temperature gradient did not exceed 6.5°C/km. The given value of the temperature gradient was considered to be the maximal possible one due to the limiting effect of convective processes on the growth of the vertical gradient.

Manabe and Wetherald found that for average cloudiness and constant relative air humidity, the doubling of carbon dioxide concentration increases the temperature at the earth's surface by 2.4°, while one-half of the present concentration decreases the air temperature by 2.3°.

In subsequent works of Manabe and Wetherald (1975) the effect of variations of CO_2 concentration on temperature at different latitudes was calculated. Using three-dimensional model of the atmospheric general circulation they found that by doubling CO_2 concentration the mean air temperature rises approximately by 3°, this value being considerably increased with increase in latitude.

Air temperature fluctuations at various latitudes for a large range of carbon dioxide concentrations were calculated by Rakipova and Vishnyakova (1973). They used a two-dimensional model of the atmospheric thermal regime which took into account the radiation heat flows, the release and consumption of heat during condensation and evaporation, the turbulent heat exchange in the atmosphere and the heat exchange between the ocean surface and its lower layers. In calculating the radiative flux, the influence of water vapor and ozone was considered.

Rakipova and Vishnyakova calculated the air temperature variations at different latitudes and altitudes as a function of the carbon dioxide concentration, which varied from zero to five times its present level.

These calculations were carried out for the cold and warm parts of the year, under conditions of clear skies and average cloudiness. They found that cloudiness, like the warm and cold periods of the year, exercised a relatively small effect on the relation between air temperatures and carbon dioxide concentrations.

It should be noted that, contrary to Manabe and Wetherald, Rakipova and Vishnyakova assumed that the absolute air humidity remains constant, which means they neglected the effect of evaporation changes with surface temperature. This, according to Manabe and Wetherald, leads to a certain underestimation of temperature fluctuations stemming from carbon dioxide concentrations.

All the foregoing investigations, of course, contain many simplifying approximations. In most cases, this problem is looked upon as a local one and does not take into account the possible changes in the general air circulation resulting from air temperature variations.

Of interest in this connection are the latest works of Manabe and Wetherald, Rakipova and Vishnyakova, which imply that average temperature variations calculated for various latitudes do not differ greatly from temperatures calculated directly on the basis of data averaged over the planet as a whole.

It is important to note that in most of the foregoing works, the feedback between the atmospheric thermal regime and the surface area of the polar ice cover was not taken into account. This effect was dealt with in part in recent work of Manabe and Wetherald, in which the change in the area of snow and ice cover is considered as a function of the atmospheric thermal regime.

As a result there was obtained a greater variation of air temperature, especially for higher latitudes, with the fluctuation of the carbon dioxide concentration in the atmosphere.

The influence of the changing polar-ice surface area on the thermal regime of the atmosphere during fluctuations of carbon dioxide was analyzed by Budyko (1972a, 1973); these results are described in Chapter 4.

This effect, it appears, is of great significance in clarifying the regularity of climatic changes during the last part of the Tertiary period and Quaternary times.

A number of works have been devoted to the study of the possible effects of atmospheric aerosol on climatic conditions. One of the first hypotheses on the relation between the atmosphere's transparency and the climate was put forward by Franklin in the eighteenth century.

Later, P. Sarasin and F. Sarasin (1901) presented the hypothesis that the Quaternary glaciations were caused by diminished atmospheric transparency during periods of increased volcanic activity. This premise was later supported by others (Fuchs, Patterson, 1947; et al.).

Humphreys dealt with this concept in greater detail in the first half of the twentieth century, claiming that after volcanic explosive eruptions, clouds of small particles form in the stratosphere, which, he said, consisted of volcanic dust. He estimated the effect of these particles on the radiation regime and concluded that they can significantly screen the radiation incident on the earth's surface, without, however, causing any appreciable effect on the long-wave radiation emitted into outer space.

According to Humphreys, the amount of particles ejected into the atmosphere over the northern hemisphere by the eruption of the Katmai volcano in 1912 was sufficient (if these particles remained in the atmosphere long enough) to bring about a drop of several degrees in the temperature at the surface. Humphreys found from observations that after intensive volcanic eruptions, temperature drops of several tenths of a degree actually took place over vast areas. He claimed that this relatively small temperature drop was occasioned by the fact that the particles of volcanic dust had remained in the atmosphere for a rather short time.

Later Humphreys' views were confirmed and elaborated by Wexler (1953, 1956), who found that the finest dust particles can remain in the stratosphere for several years after volcanic eruptions.

According to Wexler, the slow clearing from the atmosphere of the volcanic dust resulting from the major eruptions that took place at the beginning of the twentieth century caused the warming that characterized the 1920s–1930s.

An important role was attributed to volcanic activity by Brooks (1950). He compiled, on the basis of volcanic rock thicknesses, a summary of volcanic activity beginning with the Paleozoic era and continuing to our times.

Brooks advanced the opinion that the increase of volcanic activity served as one of the factors contributing to the development of continental glaciation.

Among the most recent studies of the effect on the climate of volcanic activity, mention must be made of the work published by Lamb (1970). It contains a list of explosive volcanic eruptions since 1500 which considerably reduced atmospheric transparency for short-wave radiation. Like Humphreys and Wexler, Lamb expressed the belief that volcanic eruptions constitute a major factor of changes in the climate.

This problem is studied in works by Budyko (1967, 1969a, 1972b), and the results are set forth in later chapters.

Changes in the Form of the Earth's Surface. The problem of the effects of these changes on climatic conditions was initially taken up by Lyell (1830–1833), who compared climatic conditions in two hypothetical cases for the structure of the earth's surface. These were: 1) that the polar zones are occupied by continents; and 2) that they are covered by oceans. In the first case, according to Lyell, the climate in high latitudes must be significantly warmer than in the second case. Lyell therefore came to the conclusion that the evolution of the form of the earth's surface was accompanied by strong climatic variations.

Voeikov (1902), Ramsay (1910), and Lukashevich (1915) attributed great importance of the changing topography on the climate. Voeikov thought that this contribution to climatic changes was greater than that of astronomical factors. Ramsay and Lukashevich shared the opinion that the cooling which caused glaciations occurred as a result of the rising of continents and the curtailment of the ocean surface which absorbs more solar energy than the terrain. They both reached the conclusion that, during the development of glaciations, the air temperature dropped mainly in high latitudes, with an insignificant decrease in low latitudes.

The effect on climatic conditions produced by changes in the earth's surface was studied in greater detail by Brooks (1950). He showed that the cool climate at high latitudes is caused primarily by the presence of polar ice, since it does not absorb much solar radiation due to its high albedo. Brooks also emphasized the role of ocean currents, which manifestly reduce the difference in temperature between low and high altitudes.

According to Brooks, a climate typical of most of the earth's history, with warm or hot conditions at all latitudes, prevailed when the continents were low-lying and widely separated by oceans and seas. The elevation of the continents, especially in polar regions, led to their cooling off and the development of glaciations that later expanded into the middle latitudes. Brooks also believed that the Permo-Carboniferous glaciation originated in the tropics, due mainly to the high level of continents and the change in the ocean currents in these regions.

The fact that Brooks attributed great importance to ocean currents is borne out by his contention that a slight intensification of the current in the Bering Strait could lead to the melting of much of the pack ice in the Arctic Ocean.

Albrecht (1947) was of the opinion that the elevation of meridional mountain chains, particularly the Rocky Mountains and the Andes, contributed greatly to the cooling of the climate during the Quaternary period. According to Albrecht, these mountains blocked atmospheric circulation to a great extent, decreased the amount of winter precipitation in the continental regions in the middle and high latitudes and, as a result, lowered the inflow of condensation heat.

A number of researchers have discussed the influence of changes in the relief of the ocean floor on climate (Ewing and Dunn, 1956, 1958; Rukhin, 1958). These authors expressed the opinion that the emergence of underwater mountain ridges in the northern part of the Atlantic, stretching from Scotland to Iceland and Greenland, resulted in the development of the Quaternary glaciations. This bottom relief suppressed the Gulf Stream's heat transfer and led to the cooling of the polar latitudes, where cold and dry air masses began to form. The interaction of these masses with warm and humid sea air resulted in abundant solid precipitation, which led to the development of continental glaciers.

Another problem discussed in several works was the influence on climatic changes of continental drift and the motion of the earth's poles.

The hypothesis of continental drift attracted great attention in the 1920s after the publication of Wegener's noted work. On the basis of this hypothesis Köppen and Wegener (1924) reconstructed the climates of the past. Later this assumption evoked strong objections, but today it is often referred to, in a somewhat revised form, in geological studies.

Polar motion has been the subject of much discussion in connection with paleomagnetic data (which testified to the position of magnetic poles in the past), and also in connection with ordinary paleogeographic data.

It is very difficult to clarify this problem because of the complexity of comparing paleomagnetic findings on different continents with each other and with paleogeographic data which are often contradictory.

It has been proposed that the location of the poles during the Cenozoic and the greater part of the Mesozoic eras was close to their present one, but a substantial change in this location occurred in earlier epochs. This assumption was applied by Strakhov (1960) in his reconstruction of past climates. Other authors also accept the possibility of marked polar motion in a later epoch as well.

Changes in Astronomical Elements. The French mathematician Adhèmar (1842) was the first to point out that changes in the elements of the earth's orbit could have been the cause of glaciations. Later this concept was developed by Karol' and other authors, but Milankovich (1920, 1930, 1941) subjected it to a most detailed analysis. In order to explain climatic changes during the Quaternary period, Milankovich applied the data on secular changes of three astronomical elements: the eccentricity of the earth's orbit, the obliquity of the earth's axis to the ecliptic, and the precession of the equinoxes or, in other words, the change in the date of the equinoxes in relation to the date of the closest approach of the earth to the sun. All these elements change with time due to disturbances caused by the moon and other celestial bodies on the earth's motion, the periods of these changes being: 92,000 years for eccentricity, 40,000 years for obliquity, and 21,000 years [should read 26,000 years, *Translator*] for the precession of the equinoxes.

Although the fluctuations of these elements do not affect the amount of solar radiation incident on the upper boundary of the atmosphere, they do influence the amount of total radiation received by separate latitudinal zones in various seasons.

Milankovich has developed an approximate theory of the radiational and thermal regimes of the atmosphere, from which it followed that the air temperature at the earth's surface during the warm time of the year, in middle and high latitudes, can vary by several degrees, depending on the ratio of these astronomical elements.

As a rule, this change coincided more or less with the temperature change during the cold part of the year (with a reverse sign). It is known from empirical studies, however, that the glaciation regime depends mainly on the temperature of the warm period (when the glaciers melt). The thermal conditions of the cold season affect the ice cover to a smaller degree.

By applying the said astronomical elements to studies of the Quaternary period, Milankovich found that the intervals of temperature drop during the warm seasons, in the zone where the glaciations mainly developed (60–70° latitude), corresponded to the ice ages, while those of temperature increase corresponded to the interglacial warmings.

The first comparison of Milankovich's results with Quaternary climatic conditions was made by Köppen, who asserted that these results agreed well with the history of Quaternary glaciations (Köppen and Wegener, 1924). Later, similar comparisons were made by many other scientists, whose conclusions, however, were contradictory. This evoked a lively discussion as to the influence of astronomical factors on the climate, and these arguments are still going on.

In the course of this discussion, several attempts were made to compile quantitative models of glaciation development that would allow us to estimate the changes in the size of polar ice covers as a function of the earth's position relative to the sun.

In several cases these attempts led to the conclusion that this influence is insignificant. But this apparently was due to the fact that these models did not take into consideration the feedback between the ice cover and the atmospheric thermal regime. When this was done, a definite correlation could be seen between these phenomena (Budyko and Vasischeva, 1971; Budyko, 1972b; Berger, 1973).

It should be noted that Berger and other authors (Sharaf and Budnikova, 1969; Vernekar, 1972) carried out new computations of the secular variations of the astronomical elements that influence the distribution of solar radiation in various latitudinal zones.

The probable effects on climate of fluctuations in the solar constant have been the subject of discussion in several papers. Unlike the meteorological solar constant (i.e. the flow of short-wave radiation at the outer boundary of the troposphere), which undoubtedly varies with the changing transparency of the stratosphere, the long-term variations of the astronomical solar constant (i.e. the energy emitted by the sun), cannot be considered established.

Many attempts have been made to determine short-period variations of the solar constant on the basis of astronomical observations. Several authors have come to the conclusion that these exist; the solar constant increases somewhat during periods of average solar activity and decreases during periods of low and high activity (i.e. during periods of very low or very high sun spot numbers).

Other authors (Allan, 1958; Ängström, 1969) believe that if there are any changes in the solar constant, they are within the margin of error of present-day measurements.

It is very difficult to prove any changes in the solar constant corresponding to the various climatic conditions in the past. Nevertheless, this hypothesis has been repeatedly referred to in several investigations, the best known of which are those by Simpson (1927, 1938) and Öpik (1953, 1958).

Simpson contended that the solar constant changed recurrently during the Quaternary period and that the amount of precipitation, air temperature and evaporation correlated with its behavior. Simpson was of the opinion that the warm and humid interglacial periods took place during the maximum radiation, and that the cold and dry periods occurred during its minimum.

Simpson assumed that continental glaciations originated during the periods of increasing solar radiation, when the temperature gradient between the pole and the equator increased, as did the atmospheric circulation, thereby enhancing evaporation, cloudiness and precipitation. Increased cloudiness kept the temperature from rising and glaciers began to develop in high latitudes due to the increased amount of precipitation.

When the radiation increased to its maximum, the temperature rose and, as a result, the glaciers melted. The subsequent attenuation of radiation led to a repetition of the glaciation cycle which stopped with a considerable decrease in radiation when there was not enough precipitation to maintain the glaciation process as a result of insufficient evaporation.

Öpik contended that the variations in the solar constant were caused by irregularities of solar radiation and exceeded 20%, the radiation maximum being in the Tertiary period while the minima were during the Quaternary ice ages. On the basis of his quantitative model of the thermal regime, Öpik determined the temperature variations corresponding to these fluctuations of the solar constant, and decided that this made it possible to explain the main regularities in climatic change.

Many astronomers do not share the oft-repeated opinion that the sun is a variable star (Mitchell, 1965). The question of the changing solar constant has been discussed recently at a special conference (The Solar Constant and the Earth's Atmosphere, 1975) at which there was noted the absence of sufficient data to answer the question about these changes.

Several articles have discussed the probability of changes in solar radiation reaching the earth, caused by the solar system passing through a nebula, whose matter either screens the solar radiation reaching the earth, or falls on the solar surface under gravity force thus increasing the solar radiation. This hypothesis has not won wide recognition and is rarely referred to in climatic studies.

According to a more popular hypothesis, the solar radius, in the course of the sun's evolution as a star, gradually diminished and its luminosity increased.

For example, Schwarzschild (1958) established that during the earth's history, solar luminosity could have increased 60%. This corresponds to a 1% gain in the solar constant over 80 million years. Such a change in the solar constant could noticeably have affected the climate of remote ages.

Another aspect of the influence of solar radiation on the climate is related to changes in solar activity. Many works apply available data on solar activity to the study of various atmospheric processes. It is assumed in these works that the flow of solar radiation in the ultraviolet and several other spectral bands can vary substantially, while the solar constant remains stable. The effect of these fluctuations on physical processes in the upper atmosphere is well known. However, its influence on processes in the troposphere that determine weather variations is not so clear. The assumption has been voiced that the change in solar activity, characterized by the number of sunspots, has exercised a great influence on the climate and caused the Quaternary glaciations (Huntington and Visher, 1922). It is difficult to discuss these hypotheses because there is no commonly accepted opinion on the relations between solar activity and tropospheric processes, and also because of the absence of reliable data on solar activity in the past, before solar observations began.

Internal Factors. Let us describe more precisely what we mean by external and internal climatic factors before we proceed with our study of climatic changes under constant external conditions.

We have already mentioned the external factors determining the climate, including solar radiation, the structure of the earth's surface, and the atmospheric composition, the latter depending on the interaction of the atmosphere with the lithosphere and with living organisms.

Sometimes parameters of oceanic conditions, sea and continental ice, the properties of atmospheric circulation, and even a meteorological element such as cloudiness are considered as external climatic factors.

We have repeatedly mentioned that the atmosphere's circulation, which is characterized by fields of pressure and winds as well as cloudiness, are as much internal climatic factors as the field of air temperature and precipitation. All these elements are interconnected and none of them can be considered independent of the other factors determining climatic conditions.

We must also mention among internal climatic factors the parameters of oceanic conditions and of the ice cover over land and water reservoirs.

It is obvious that the genesis of the climate, in a broad sense, embraces not only atmospheric but also physical phenomena in the hydrosphere and the entire complex of hydrometeorological events over the land surface, including the development of glaciations, if any.

Thus, the relief structure which determines the surface area and the depth of seas and oceans, as well as the surface area and altitudes of continents, should be considered a climatic factor, and is called the underlying surface. Changes in the underlying surface, for instance, glaciations, must be considered among the climatic elements which are determined by the same external factors (incident radiation, relief structure and atmospheric composition), although in particular cases they can be looked upon as factors of climatic genesis at a given period.

When considering the atmosphere-ocean-ice cover system, we must ask whether climatic conditions in this system can change when external climatic factors are constant.

Lorenz (1968) studied this problem by analyzing equations of the theory of climate. He showed that two solutions of these equations were possible. The first solution gives one type of stable climate which is obtained as a result of averaging instantaneous values of fields of meteorological elements over a long-range period. This type of climate, called "transitive" by Lorenz, is single-valued under the given external conditions.

The second solution presents several types of a stable climate for the same external conditions ("intransitive" climate). In this case the climate can change significantly without any visible external causes.

This subject is further elaborated in the next chapter. Lorenz has expressed the interesting idea that the climate in our times can be quasi-intransitive. This means that a climate which is determined over a long but finite time interval can depend on the initial conditions and therefore be different for various time periods, under constant external factors. Whether the present-day climate is quasi-intransitive or not is still open to question.

The possibility of auto-oscillatory processes developing in the atmosphere-ocean-ice cover system is also of vital importance in understanding climatic changes. The simplest example of such an auto-oscillatory system is presented by the level of oscillations in a drainless reservoir. As shown by the authors, and Yidin (1960), the level of such a reservoir cannot be stable even in the absence of climatic changes, if its shores are vertical, i.e. if its surface area does not depend on the level. The reason for this instability is that the sum of random annual oscillations of the components of the reservoir's water balance increases indefinitely with time.

Since the shores of natural reservoirs are sloped, the dependence of the reservoir's surface on its level serves as a stabilizing factor on the oscillations. In this work it was established that if the successive annual oscillations of the Caspian Sea level were mutually independent, the total amplitude of these oscillations would be equal to the sum of random annual level oscillations. In this case they would be an example of a pure auto-oscillatory process.

Drozodov and Pokrosvkaya (1961) showed that the successive annual oscillations of the Caspian Sea level are not independent. A study of their interaction lead to the conclusion that the main part of the amplitude was caused by external factors (i.e. by general climatic fluctuations), a significant part of this amplitude being a consequence of an auto-oscillatory process.

Far more complicated auto-oscillatory processes can develop in the atmosphere-ocean-polar ice system. Models of such processes were first analyzed by Shuleikin (1941, 1953).

The possibility of explaining Quaternary climatic changes as auto-oscillatory processes has been discussed in several papers including the above works of Plass and of Wilson (1964). The latter's work was elaborated by Flohn (1969) who explained the Quaternary ice ages by an instability of the Antarctic glacier that caused auto-oscillatory processes in the atmosphere.

Drozdov and Grigor'eva (1971) adopted the same approach to the study of cyclic fluctuations of atmospheric precipitation over the territory of the U.S.S.R. They came to the conclusion that these fluctuations were of an auto-oscillatory nature. By using empirical relations established in this and preceding works, it was possible to predict the expected precipitation regime for the next decade for various regions for the U.S.S.R.

The Present Stage of the Problem. Aside from the propositions outlined above, there are several other hypotheses which, for various reasons, are less frequently referred to in climatic studies. However, those we have gone into comprise an adequate list, illustrating the wide range of opinions on the causes of climatic changes.

All contemporary reviews note the absence of a generally accepted theory on this problem (Schwarzbach, 1950, 1961, 1968; Sellers, 1965; Sinitsyn, 1967; Mitchell, 1968; Shell, 1971).

We think this is inevitable at a stage when many aspects of the problem are dealt with mostly in qualitative terms, without any reliable quantitative calculations.

It may be assumed that a precise estimate of the problem will be reached only after general theories have been developed on climate and on the evolution of the atmosphere, ocean and upper layers of the lithosphere, and that of the sun. Since it may take several decades to adequately develop these theories, it becomes urgent at the present time to develop approximate methods for studying climatic changes which will enable us to obtain quantitative estimates of the effects of these factors that so obviously influence the climate. If these hypotheses succeed in explaining the chief regularities observed in climate, there will be no need to apply additional propositions for explaining the processes under study.

An urgent problem today is to work out a quantitative method not only for explaining past climates but primarily to predict climatic changes in the next decades which may transpire as a result of man's economic activity.

There are several ways of solving this problem. One of them is based on data concerning the energy and water balance of the earth. Such a method was developed by the author and his colleagues during the past several years. The main results obtained by these specialists are discussed in the following chapters.

2 The Genesis of the Climate

2.1 The Energy Balance of the Earth

In Chapter 1 we noted that the climate is determined by the solar radiation incident on the upper boundary of the atmosphere, by the atmospheric composition, and by the structure of the earth's surface. In order to study the processes that go on under the influence of these factors, or, in other words, in order to explain the genesis of the climate, data on the energy balance of the earth are very important. These data clarify the transformation of solar energy in the atmosphere, in the ocean, and on the earth's surface.

Components of the Energy Balance. The problem of studying the energy (or thermal) balance of the earth was formulated in the 1880s by A. I. Voeikov, the outstanding Russian climatologist. Later important contributions to this study were made by W. Schmidt, A. Ångström, H. Sverdrup and F. Albrecht.

Since 1945 systematic studies of the earth's energy balance have been conducted at the A. I. Voeikov Main Geophysical Observatory (Leningrad), where two world atlases were compiled on the energy balance (Atlas of Heat Balance, 1955; Atlas of Heat Balance of the World, 1963), and other data obtained on the geographic distribution of the heat balance components (Budyko, 1956, 1971).

These efforts provided information on all the components of the heat balance equation of the earth's surface:

$$R = LE + P + A, \qquad (2.1)$$

where R is the radiation balance of the earth's surface, LE is the heat expenditure for evaporation (L is the latent heat of evaporation, E is evaporation); P is the turbulent heat flow from the underlying surface to the atmosphere; A is the heat flow from the underlying surface to the deeper layers.

Alongside with these data on the heat balance of the earth's surface, information was obtained on the components of the heat balance of the earth-atmosphere system, i.e. the heat balance of a vertical column of a unit cross-section that passes through the atmosphere and the ocean (or, on a continent, through the atmosphere and the upper layer of the lithosphere).

The heat balance equation for the earth-atmosphere system is

$$R_s = F_s + L(E - r) + B_s, \qquad (2.2)$$

where R_s is the radiation balance of the earth-atmosphere system, and F_s is the difference between the input and loss of heat in this column, as a result of horizontal motions in the stratosphere and hydrosphere. This quantity is equal to the sum of the heat input to the atmosphere (F_a) and to the hydrosphere (F_0); Lr is the heat input from condensation, which is considered proportional to the amount of precipitation r, and B_s is the change of heat content in a vertical column passing through the earth-atmosphere system, caused mainly by the heating and cooling of the ocean's upper layers.

The radiation balance of the earth's surface (R) is equal to the difference between the short-wave radiation absorbed by the earth's surface, and its effective long-wave emission, while the radiation balance of the earth-atmosphere system (R_s) is equal to the difference between the solar radiation absorbed by this system and the long-wave radiation at the upper boundary of the atmosphere being emitted into outer space. The values of R and R_s in equations (2.1) and (2.2) are considered positive when they correspond to heat gain, while all the other terms in these equations are positive in the case of heat loss.

When studying the energy balance, it is necessary to take into consideration the water balance of the earth, the components of which form the following equation:

$$r = E + f + b, \qquad (2.3)$$

where f is the run off for dry land, or the horizontal redistribution of water for bodies of water, and b is the change of moisture content in the lithosphere's upper layers, or the change of water content in water bodies due to level oscillations.

In this equation r is always positive, E and f positive when they indicate water loss, b positive when the amount of water in the soil or in bodies of water increases.

We shall present the earth's heat balance components calculated from the data offered by the "Atlas of Heat Balance of the World" (1963).

On the basis of these maps, the mean latitudinal values of the heat balance components have been calculated for the surfaces of continents, oceans and the earth as a whole. These results are presented in Table 3.

It follows from this table that only one element of the heat balance, i.e. the radiation balance, has a similar distribution for land and oceans.

Table 3
Mean Latitudinal Values of the Heat Balance Components ($kcal/cm^2 \cdot yr$)

Latitude	Oceans				Continents			Earth as a Whole			
	R	LE	P	F_0	R	LE	P	R	LE	P	F_0
70–60°N	23	33	16	−26	20	14	6	21	20	9	−8
60–50	29	39	16	−26	30	19	11	30	28	13	−11
50–40	51	53	14	−16	45	24	21	48	38	17	−7
40–30	83	86	13	−16	60	23	37	73	59	23	−9
30–20	113	105	9	−1	69	20	49	96	73	24	−1
20–10	119	99	6	14	71	29	42	106	81	15	10
10–0	115	80	4	31	72	48	24	105	72	9	24
0–10°S	115	84	4	27	72	50	22	105	76	8	21
10–20	113	104	5	4	73	41	32	104	90	11	3
20–30	101	100	7	−6	70	28	42	94	83	15	−4
30–40	82	80	9	−7	62	28	34	80	74	12	−6
40–50	57	55	9	−7	41	21	20	56	53	9	−6
50–60	28	31	8	−11	31	20	11	28	31	8	−11
Total	82	74	8	0	49	25	24	72	60	12	0

The maximum net radiation balance, both for continents and oceans, occurs in the tropics, the average latitudinal values of the radiation balance being almost constant within the tropical zone.

The difference between the radiation balance values for land and ocean varies very markedly with latitude. This difference is greatest in the tropical zone, where it comprises one-third of the ocean's radiation balance. The radiation balances for land and oceans converge with latitude and at 50–70° become practically equal. One of the reasons for this is the increase of the mean albedo of the water surface with latitude.

The average latitudinal values of heat expenditure on evaporation for the continents have a main maximum at the equator, while in the latitudes of the high-pressure belt, the evaporation decreases. The evaporation increases slightly with latitude in both hemispheres and more noticeably in the northern hemisphere. This is due to the greater amount of precipitation in mid-latitudes than in the dry zones of lower latitudes. With further increase of latitude, the evaporation decreases due to insufficient heat.

In contrast to continental conditions, the maximum heat expenditure for evaporation for oceans, averaged over the latitudes, takes place in the high-pressure belts. It should be noted that at the latitudes of 50–

70°, where the radiation balances of the continents and oceans are approximately the same, the heat expenditure for evaporation over the ocean is much higher than over the continents. This is partly explained by the fact that evaporation from the ocean consumes much heat supplied by ocean currents.

The turbulent heat flow in the oceans increases monotonically with latitude. On the continents it reaches its maximum in the high-pressure zones; it is somewhat lower at the equator and drops drastically at high latitudes. The totally different character of these changes on land and on sea indicates that the transformation of air masses over continents and oceans is caused by different mechanisms.

The distribution of average latitudinal values of incoming and outgoing heat in the oceans shows that the ocean currents carry the heat mainly from the zone between 20°N and 20°S, the maximum of the heat loss being somewhat shifted to the north of the equator. This heat is transferred to higher latitudes and is dissipated mainly in the latitudes of 50–70°N where particularly strong warm currents prevail.

The latitudinal distribution of the components of the heat balance for the earth as a whole is characterized by a regularity which depends on the ratio between the surface areas of continents and oceans in various zones. The average values of the heat balance components for continents and oceans are given in Table 4.

It is evident from this table that on three of the six continents (Europe, North America and South America) most of the net heating is consumed by evaporation. The other three continents (Asia, Africa, and Australia) show the reverse, indicating the prevalence of dry climate. The mean parameters of the heat balance differ very little on all three oceans. It should be noted that the sum of the heat expenditure and the turbulent heat exchange for each ocean is very close to the net radiation

Table 4
Heat Balance Components for Continents and Oceans ($kcal/cm^2 \, yr$)

Component	Continent						Ocean		
	Europe	Asia	Africa	N. America	S. America	Australia	Atlantic	Pacific	Indian
R	39	47	68	40	70	70	82	86	85
LE	24	22	26	23	45	22	72	78	77
P	15	25	42	17	25	48	8	8	7

balance value. This means that the heat exchange between the oceans caused by ocean currents does not affect the heat balance of each ocean, taken separately. A small excess of the radiation balance over the total of the turbulent heat flux and the expenditure for evaporative heating in the Atlantic corresponds to the transfer of a certain amount of heat from the Atlantic to the Arctic Ocean. Since this difference is within the margin of error, further study of this problem is clearly necessary.

Table 4 does not include data on the heat balance of the Antarctic continent or the Arctic Ocean, since this information is still unreliable.

On the basis of the foregoing data and available estimates of heat balance components for the polar region, calculations have been made of the heat balance components for all continents, oceans, and the earth as a whole. The results are given in the last line of Table 3.

From these data it follows that over 90% of the radiation balance on the ocean surface is consumed by evaporation and only 10% by direct turbulent heating of the atmosphere. On the continents these expenditures are almost equal. For the entire earth the heat expenditure comprises 83% of the radiation balance while the turbulent heat exchange comprises 17%.

The Atlas of Heat Balance of the World (1963) provides data on average latitudinal values of the heat balance components for the earth-atmosphere system and the atmosphere (Budyko, 1971). It can be seen that the relation between the heat balance components for the earth-atmosphere system in various latitudes, differ substantially for average annual conditions.

In addition to the large inflow of radiative energy into the equatorial zone, there is a significant heat influx in the earth-atmosphere system as a result of water phase transformations, i.e. due to the difference between heat of condensation and heat energy expenditure for evaporation. These heat sources provide a great amount of the energy that is consumed by atmospheric and ocean advection, for which a comparatively narrow near-equatorial zone is an extremely important energy source.

In higher latitudes, such as about 30–40°, there is a radiation surplus (positive balance) of the earth-atmosphere system which decreases with latitude and the heat expenditure on water phase transformations reaches a significant value. In the greater part of this zone, the heat expended on phase transitions is comparable with the radiation surplus; therefore the heat redistribution by atmospheric and ocean currents is comparatively small.

The zone of radiation deficit (negative balance), which increases in absolute value with latitude, is located at latitudes above 40°. The radiation deficit in this zone is compensated by the inflow of heat transferred by atmospheric and ocean currents, the ratio of heat balance components that compensate for the energy deficit being different at various latitudinal zones. In the 40–60° zone, the main source of energy is the energy released during water vapor condensation in excess of the energy expenditure for surface evaporation.

At these latitudes the inflow of heat redistributed by ocean currents also plays an important role. The redistribution of heat by atmospheric circulation becomes the main source of thermal energy at higher latitudes, and especially in the polar region, where the heat inflow from condensation is small and the effects of ocean currents is either absent (south polar zone), or is weakened by a permanent ice cover (north pole zone).

Calculations show that the zonally-averaged values of the atmosphere's radiation balance change less than other components of the heat balance. The large deficit observed at all latitudes is compensated mainly by the heat inflow from condensation.

The heat inflow from the earth's surface caused by turbulent heat exchange plays a smaller part, although its influence on the atmospheric heat balance is significant.

Mean latitudinal values of the heat balance components for the earth-atmosphere system for two half-years are shown in Table 5, where Q_a is the radiation absorbed in the earth-atmosphere system, and I_s is the outgoing long-wave radiation at the upper boundary of the atmosphere. The difference between these two values is equal to the radiation balance R_s of the earth-atmosphere system. The total heat redistributions caused by horizontal motions in the atmosphere is equal to $F_0 + L(E - r)$ and is designated by C_a. The values of F_0 in Tables 3 and 5 were determined by different methods and vary slightly.

The left-hand segment of the upper half of the table and right-hand segment of the lower half refer to the warm half-year, while the other two segments refer to the cold period.

As shown by Table 5, the absorbed energy is not the only factor that governs the amount of outgoing radiation. In middle and high latitudes during the cold period (and in high southern latitudes throughout the year), the main source of heat is its transport from lower latitudes by atmospheric circulation.

The chief component of the two heat-exchange processes in the ocean consists of seasonal heat accumulation and discharge by the mass of ocean water. At some latitudes this component of the heat balance reaches 25–30% of the outgoing radiation. The heat redistribution by ocean currents plays a smaller part, although in certain zones it may reach 15–20% of the radiation emitted into space. It should be noted that the values of these components of the ocean heat balance refer to the

Table 5
Mean Latitudinal Values of the Heat Balance Components of the Earth-Atmosphere System for Two Half-Years ($kcal/cm^2 \cdot month$)

Latitude	First Half-Year (IV-IX)					Second Half-Year (X-III)				
	Q_a	C_a	F_0	B_s	I_s	Q_a	C_a	F_0	B_s	I_s
80–90°N	7.8	−4.5	0	0.8	11.5	0.1	−9.1	0	−0.8	10.0
70–80	8.2	−4.4	0	0.8	11.8	0.5	−9.3	0	−0.8	10.0
60–70	11.5	−1.8	−0.4	1.2	12.5	1.8	−6.7	−1.5	−1.2	11.0
50–60	14.6	0.0	−1.3	2.8	13.1	4.0	−4.7	−0.5	−2.8	12.0
40–50	16.9	1.0	−2.0	3.9	14.0	6.5	−2.9	0.6	−3.9	12.0
30–40	19.2	1.4	−1.7	4.3	15.2	9.5	−0.9	0.7	−4.3	14.0
20–30	20.0	2.0	−0.4	2.9	15.5	13.7	0.9	0.6	−2.9	15.0
10–20	19.7	2.4	0.9	1.4	15.0	17.0	1.8	1.0	−1.4	15.0
0–10	18.4	1.4	2.2	−0.1	14.9	18.7	2.1	1.4	0.1	15.0
0–10°S	18.0	2.3	2.2	−1.5	15.0	19.7	2.4	0.8	1.5	15.0
10–20	16.2	2.6	0.9	−2.5	15.2	20.7	3.5	−0.3	2.5	15.0
20–30	13.0	1.4	0.3	−3.4	14.7	20.6	3.5	−1.2	3.4	14.0
30–40	8.9	−0.4	−0.4	−4.2	13.9	18.9	2.0	−1.4	4.2	14.0
40–50	5.9	−1.9	−1.6	−3.5	12.9	15.9	0.0	−0.6	3.5	13.0
50–60	3.3	−3.9	−2.6	−2.5	12.3	12.8	−2.6	0.6	2.5	12.0
60–70	1.0	−9.5	0	−0.8	11.3	8.1	−4.3	0	0.8	11.0
70–80	0.2	−9.8	0	0	10.0	4.4	−6.5	0	0	10.0
80–90	0.0	−8.8	0	0	8.8	3.4	−6.9	0	0	10.0

total surface of the latitudinal zones. In most cases this reduces their values significantly as compared to the heat balance components for the surface areas of the ocean.

Table 5 indicates that in order to determine the outgoing radiation from other components of the heat balance, it is necessary to take into consideration all the components included in the table. The values of the heat balance components for the earth as a whole are presented schematically in Fig. 1 (Budyko, 1971).

The total flux of solar radiation reaching the outer boundary of the troposphere is about 1000 $kcal/cm^2 \cdot yr$. Due to the spherical form of the earth, a unit area of the upper boundary of the troposphere receives, on the average, a quarter of this amount, i.e. about 250 $kcal/cm^2 \cdot yr$. Assuming the earth's albedo as $\alpha_s = 0.33$, we find that the short-wave radiation absorbed by the earth is equal to 167 $kcal/cm^2 \cdot yr$. This quantity is depicted in Fig. 1 by the arrow $Q_s(1 - \alpha_s)$.

According to the foregoing data, the earth's surface receives 126 $kcal/cm^2 \cdot yr$. of short-wave radiation. The mean albedo of the earth's surface α (considering the different amount of incident solar radiation

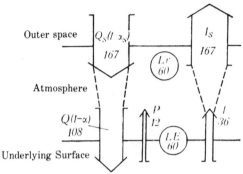

Fig. 1. Heat balance of the earth (components are given in terms of kcal/cm²·yr).

for various regions) is equal to 0.14. Thus the earth's surface absorbs 108 $kcal/cm^2 \cdot yr$, which is indicated in Fig. 1 by arrow $Q(1-\alpha)$, while it reflects 18 $kcal/cm^2 \cdot yr$.

We see, therefore, that the atmosphere absorbs 59 $kcal/cm^2 \cdot yr$, or a great deal less than the surface.

The radiation balance of the earth's surface is thus equal to 72 $kcal/cm^2 \cdot yr$, while the effective radiation at the earth's surface is, on the average, equal to 36 $kcal/cm^2 \cdot yr$ (arrow I). The total amount of the long-wave radiation of the earth is equal to the amount of absorbed radiation and is about 167 $kcal/cm^2 \cdot yr$ (arrow I_s).

It should be noted that the ratio of the effective radiation from the earth's surface to the total radiation of the earth (I/I_s) is considerably less than the corresponding ratio of the absorbed radiations $Q(1-\alpha)/Q_s(1-\alpha_s)$. This fact indicates the strong contribution of the "greenhouse effect" to the thermal regime of the earth.

Due to the "greenhouse effect" the earth's surface receives about 72 $kcal/cm^2 \cdot yr$ of radiative energy (radiation balance), which is partly consumed by water vapor (60 $kcal/cm^2 \cdot yr$, shown by circle LE), and partly returned to the atmosphere in the form of turbulent heat flow (12 $kcal/cm^2 \cdot yr$, arrow P). As a result, the heat balance of the atmosphere is comprised of the following elements:

1) incoming heat from the absorbed short-wave radiation equal to 59 $kcal/cm^2 \cdot yr$,

2) incoming heat from water vapor condensation equal to 60 $kcal/cm^2 \cdot yr$ (in Fig. 1 circle LE);

3) incoming heat from the turbulent heat transfer from the earth's surface equal to 12 $kcal/cm^2 \cdot yr$;

4) expenditure of heat by effective radiation into outer space equal to $I_s - I$, i.e. 131 $kcal/cm^2 \cdot yr$.

2 THE GENESIS OF THE CLIMATE

The last term is equal to the sum of the first three components of the heat balance.

Water Balance. Let us consider the water balance of the continents first. The total precipitation over land has been measured over a long period of time at numerous meteorological stations. Therefore there are no difficulties in compiling precipitation maps based on observational data.

The average annual amounts of precipitation on continents determined from the precipitation maps compiled by Kuznetsova and Sharova (1964) are given in Table 6. The table does not contain data for the Antarctic, but these were taken into consideration in determining the average precipitation for the entire land area.

It should be noted that meteorological instruments often give an inadequate estimate of the amount of precipitation, particularly of snow which is often blown away by the wind.

Measurements of precipitation by various instruments contain different systematic errors. In compiling precipitation maps, attempts have been made to correct most of these errors. Generally, however, the attempted corrections have proved insufficient. Recent investigations show that in the regions subject to heavy snowfalls, the actual precipitation can exceed previously accepted figures by a high percentage. The amount of rainfall is measured, as a rule, with a smaller margin of systematic error. That is why the figures in Table 6 are somewhat underestimated.

In the absence of large-scale observations, the average evaporation from land surfaces can be determined solely by calculations. Table 6 gives the values estimated on the basis of the pertinent maps in the Atlas of Heat Balance of the World (1963).

It follows from the table that the ratio of evaporation to precipitation varies greatly for different continents. This ratio is close to unity in

Table 6
Precipitation and Evaporation on Continents

Continent	Precipitation (cm/yr)	Evaporation (cm/yr)
Europe	64	40
Asia	60	37
N. America	66	38
S. America	163	75
Africa	69	43
Australia	47	37
Entire terrain	73	42

Australia, whereas in South America the evaporation comprises less than half of the precipitation.

The components of the ocean water balance can be calculated on the basis of the world precipitation map compiled by Kuznetsova and Sharova (1964), and the world evaporation map from the Atlas of Heat Balance of the World (1963) (Table 7).

The difference between evaporation from the ocean surface and precipitation is equal to the river runoff from the continents into the oceans. For a given ocean, this difference is equal to the sum of river runoff and the horizontal water transfer between oceans due to circulation processes. It is difficult to determine this quantity directly because the difference between the inflow and outflow of water is comparatively small, and each quantity is determined with a substantial error. It is simpler to evaluate the water exchange between the oceans as the remainder term of the water balance of every ocean, although in this case, there is very little precision in the corresponding values as well.

These values of river runoff for every ocean are given in Table 7 according to Zubenok (1956).

This table does not include data on the water balance of the Arctic Ocean, which are less precise than for the other oceans, but they are taken into account in determining the water balance of the earth as a whole.

It follows from the table that the sum of precipitation and river runoff is less than the evaporation in the Atlantic Ocean. This deficit is compensated by water flowing in from other oceans including the Arctic, where evaporation is considerably smaller than the sum of precipitation and river runoff. In the Indian Ocean this sum is slightly lower than the evaporation, while in the Pacific it is higher, corresponding to the transfer of surplus water to other oceans.

On the basis of precipitation and evaporation data given above for continents and oceans, we find that for the earth as a whole, the annual amount of precipitation comes to 102 cm, which is equal to the amount of evaporation. This figure is noticeably higher than similar values deter-

Table 7
Water Balance of the Oceans

Ocean	Precipitation (cm/yr)	Evaporation (cm/yr)	River Discharge (cm/yr)
Atlantic	89	124	23
Pacific	133	132	7
Indian	117	132	8
World ocean	114	126	12

mined in the course of most of the earlier studies (Wüst, 1936, Möller, 1951), but for the reasons enumerated above, we must assume that it is somewhat underestimated.

The precipitation maps of Kuznetsova and Sharova (1964) and the evaporation maps from the Atlas of Heat Balance of the World enable us to determine the water balance of the earth's latitude zones.

The dependence on latitude of the water balance components is illustrated by Fig. 2. We find in various zones that the influx of water vapor into the atmosphere from evaporation can be either greater or smaller than its consumption by precipitation, the main source of water vapor in the atmosphere being chiefly the high pressure zones, where evaporation predominates over precipitation. The expenditure of this water vapor surplus occurs in the near equatorial zone as well as in middle and high latitudes, where precipitation exceeds evaporation. It is obvious that the quantity f, which is equal to the difference between precipitation and evaporation, is at the same time equal to the difference between the input and outflow of water vapor in the atmosphere.

Reliability of Data on Components of the Energy Balance. The total amount of data on the energy balance components obtained during the last decades is comparable to that of other meteorological elements observed at the basic network of weather stations. The precision of these data is of importance in the study of climate.

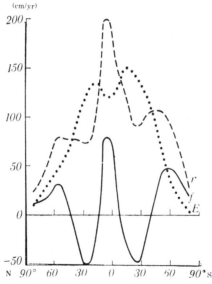

Fig. 2. Mean latitudinal distribution of water balance components of the earth.

Most of the maps of the heat balance components are compiled by means of calculations as there are very few, if any, observational data available.

In several cases, after these maps were published, direct observations of these components were collated. Thus the calculated maps served as a preliminary estimate of anticipated data from observations.

The first work of this kind was to determine the average values of the radiation balance of continental surfaces. The first maps of this parameter were published in the early half of the 1950s, when long-term observational data of the radiation balance did not exist.

In the second half of the 1950s, during the International Geophysical Year, observations of this type were initiated in many parts of the world. Their results were first compared with previous calculations by Robinson (1964), who found a good correlation between the measured and calculated data.

Later, studies of this kind were frequently made on the basis of a much larger amount of data. Figure 3 gives a comparison of the measured (R_m) and calculated (R_c) average monthly values of the radiation balance for 27 stations situated on different continents. The correlation coefficient between these values in this case is equal to 0.98.

The measurements of the radiation balance on the ocean commenced later than on the continent, and the great amount of material

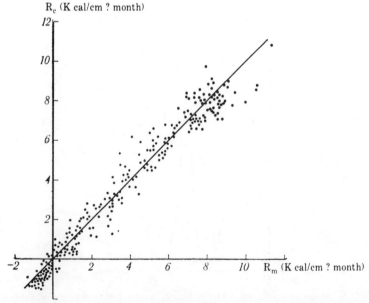

Fig. 3. Comparison of the measured (R_m) and calculated (R_c) average monthly values of the radiation balance of the continents.

accumulated only by the second half of the 1960s. Therefore the previously published maps served as a prognosis of what could be expected in the future. A comparison of the measured and calculated values of the radiation balance of ocean surfaces was carried out by Strokina (1967), and the results of this comparison are presented in Fig. 4. The correlation coefficient in this case is equal to 0.96.

Another example of verification of the calculated maps is by means of the radiation balance of the earth-atmosphere system. It is known that these maps were compiled before the first direct measurements were carried out by means of artificial satellites. A comparison shows that although this correlation is worse than for similar maps of the earth's surface, the main regularities established by direct observations correspond to those of the calculated maps.

Note that the distribution of the heat balance components of the earth-atmosphere system reveals an interesting peculiarity in desert regions. It appears from the calculated map published in the Atlas of Heat Balance of the World (1963), that the annual radiation balance of the Sahara Desert is close to zero, which means that there is no source of heat advection in this region for other regions. This paradoxical result was later confirmed by direct measurements (Raschke, Möller, and Bandeen, 1968).

The foregoing examples comprise only a small portion of the available material on heat balance components (Budyko, 1971); analysis proves that these data are quite reliable for the study of climatic genesis.

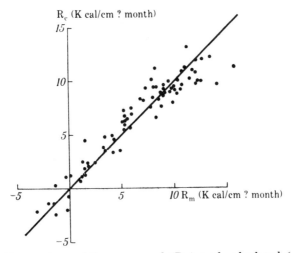

Fig. 4. Comparison of the measured (R_m) and calculated (R_c) average monthly values of the radiation balance of the oceans.

2.2 Semiempirical Theory of Climate

Contemporary Climate Theories. The theory of climate aims to determine, by means of physical deduction, the average distribution of meteorological elements in space and time and their variability under the influence of given external factors. In order to develop a theory of climate, methods of fluid mechanics and thermodynamics are applied, as well as the laws of energy transport and phase transformations of water, along with other formulations of contemporary physical meteorology. In a review recently published (Inadvertent Climate Modification, 1971), F. Thompson divided the existing theories into four groups:

1) models referring to the planet as a whole, describing the average meteorological conditions for the entire atmosphere;

2) models for zonally-averaged conditions in which the longitudinal variation of the fields is not considered; in these models large-scale atmospheric motions are statistically parameterized;

3) semiempirical models based on empirical regularities established by observations;

4) general models taking into account all the details of large-scale atmospheric motions.

This classification is not strictly logical because these groups are not uniform as regards two criteria: the range of application and the use of empirical relations. Nevertheless, it reflects the main trends of modern attempts in the development of the theory of climate. One of the first numerical models of this theory was developed by Milankovich (1920, 1930), who calculated the distribution of average latitudinal temperatures in the atmosphere based on the amount of radiation incident on the upper boundary of the atmosphere. Milankovich believed that the thermal regime is determined by the heat exchange in each separate latitude zone. He did not take into account the greenhouse effect of the atmosphere, nor the variations of albedo with latitude. Milankovich also neglected the effect on the thermal regime of the meridional heat transfer in the atmosphere, as well as the heat exchange caused by the phase transformation of water and the interaction between the atmosphere and the ocean. Significant errors caused by these factors were compensated for by the selection of certain special parameters, and as a result Milankovich obtained a distribution of average latitudinal temperature that does not differ greatly from the observed one. This correlation, however, was highly provisional.

The first step in the application of methods of dynamic meteorology to the development of the theory of climate was taken by Kochin (1936). Kochin calculated the average meridional profiles of pressure and wind velocity in the lower atmosphere, on the basis of the given distribution of air temperature and pressure at the earth's surface. Later, Dorodnitsyn, Izvekov and Shwets (1939) applied the same method to a model of summer zonal circulation in the northern hemisphere. This trend in research was further developed by Blinova (1947), who calculated the average distribution of temperature, pressure, and wind velocity in the northern hemisphere.

Of particular importance in the development of the theory of climate are the works of Shuleikin (1941), who established that interaction between the ocean and the atmosphere was a major and frequently dominant factor in the formation of climatic conditions.

The treatment of the physical mechanism of heat exchange and moisture exchange in the models has posed great difficulties. Several works by Rakipova (1967) were devoted to this problem; she calculated the distribution of average latitudinal values of air temperature and other elements by considering all the main forms of heat input in the atmosphere which contribute to the heat balance.

Later, many new models were developed making it possible to calculate the average fields of meteorological elements. Progress in computational techniques facilitated the development of quantitative models representing unaveraged fields of meteorological elements. Calculations of these fields for a sufficiently long period of time resulted in finding average fields that characterize climatic conditions.

Numerous studies in this direction were carried out by Smagorinsky (1963), Smagorinsky, Manabe, and Holloway (1965), Mintz (1965), Manabe and Bryan (1969), Holloway and Manabe (1971), Wetherald and Manabe (1975).

In the abovementioned works, Manabe and Bryan developed a numerical model of the theory of climate that takes into account the effect of oceanic circulation on climatic conditions. Holloway and Manabe were the first to compile theoretical world maps of the main components of the heat and water balance of the earth's surface. These maps correlated well with similar maps based on empirical data.

An original approach to the problem of the general theory of climate was suggested by Shvets and his colleagues (1970). The improved method of parameterization of the condensation and radiation processes introduced in this model allowed them to take into account the direct and feedback relations between these processed and the dynamic state of the atmosphere, and thus all the links in the chain of a general circulation model were joined.

Such models must meet a number of requirements in order to be applied to the study of climatic changes. These requirements are not as essential for models of the contemporary climatic regime. The first requirement is that the model must not contain empirical data on the distribution of climatic elements, especially those that are highly sensitive to the effect of climatic changes. The second requirement is that the model must realistically take into account all types of heat inflow which tangibly affect the temperature field and must conform with the energy conservation law. The third requirement is that the model must consider the main direct and feedback relations of various climatic elements.

The first two requirements are more or less obvious, but the third one must be elaborated on.

The stability of climate is achieved by those feedback relations between its elements that can be called negative (Inadvertent Climate Modification, 1971). This feedback attenuates perturbations in meteorological fields and helps them to revert to their climatic norms. An example of negative feedback is the dependence of long-wave radiation on the earth's surface temperature. When the temperature rises, radiation increases, resulting in growth of heat energy expenditure and causing a decrease in temperature use. Another example of negative feedback is the dependence of atmospheric heat transfer upon the air temperature gradient. As a rule, the heat flow in the atmosphere is directed from regions of higher temperature toward regions of lower temperature, which leads to a leveling of the temperature distribution.

Positive feedback, on the other hand, plays an important part in climatic changes contributing to an increase in perturbations of meteorological elements, thus decreasing climate stability.

An example of positive feedback is the dependence of absolute humidity upon its temperature mentioned in Chapter 1. As temperature rises, evaporation from water or wet surfaces increases, thus leading to a comparatively stable relative humidity in most climatic regions (with the exception of dry continental areas). Under these conditions, the absolute air humidity increases with temperature, as confirmed by numerous empirical data.

Since the effective long-wave radiation decreases with the growth of absolute humidity, the latter facilitates a further temperature rise. It is obviously necessary to take this feedback into account in numerical models of the thermal regime used for studying climatic changes.

Another positive feedback — the influence of snow and ice sheets on the earth's surface albedo — plays an even greater role in the laws governing the changes in the atmosphere's thermal regime.

Influence of Snow and Ice Covers on Climate. The first to study the influence of the snow cover on climatic conditions was Voeikov (1884), who proved that a snow cover aids in reduction of the air temperature over its surface. Later, Brooks (1950) came to the conclusion that the ice cover, due to its high albedo, significantly reduces the air temperature, as a result of which climatic fluctuations increase greatly following the formation and melting of ice covers.

The values of the albedo of the earth's surface and of the earth-atmosphere system in high latitudes can be determined on the basis of data obtained during expeditions to the Arctic and Antarctic as well as from satellite observations; these figures may be compared with the albedo of regions without any snow or ice cover.

Available data indicate that during the summer months the albedo of the ice surface in the Central Arctic is about 0.70 and is about 0.80–0.85 in the Antarctic.

If we recall that the mean albedo of the earth's surface in regions free of snow and ice is about 0.15, we come to the conclusion that the snow and ice cover results in a manifold reduction of radiation absorbed by the earth's surface, other conditions being equal.

Snow and ice also greatly affect the albedo of the earth-atmosphere system. According to recent satellite data (Raschke, et al., 1968, 1973), the system's summer albedo in the central Arctic and Antarctic is approximately 0.60, which is about twice that of the earth's albedo as a whole.

It is obvious that such great differences in albedo values must strongly affect the thermal regime of the atmosphere.

If a snow or ice cover is formed on the surface as a result of an air temperature drop, this must facilitate a further temperature decrease as well as an expansion of the snow and ice covers. The reverse is true when the temperature rises, if it leads to melting snow and ice.

That this feedback greatly affects the air temperature distribution near the earth's surface is proved by taking this feedback into account in a numerical model of the atmosphere's thermal regime (Budyko, 1968). A simple example may be given in order to estimate this influence indicating how the mean global air temperature will change if the earth's surface under cloudless conditions is completely covered with snow and ice (Budyko, 1962, 1971).

Let us consider a hypothetical case in which the earth's surface is completely covered by ice and snow, and clouds are absent. Under these conditions the earth's albedo is much higher than the actual value and this should cause a change in the mean global temperature. The effective temperature of the earth corresponding to its long-wave radiation is proportional to $(1 - \alpha_s)^{1/4}$, where α_s is albedo. Therefore the absolute effective temperature when the albedo changes from α_s' to α_s'' will change as $[(1 - \alpha_s'')/(1 - \alpha_s')]^{1/4}$. If we take the actual albedo of the earth as equal to 0.33 and the albedo of dry snow surface as 0.80, we find that for the earth covered with snow, the mean effective temperature will drop by approximately 75°C.

The mean air temperature near the ice-covered surface will probably drop even more. At present, the mean temperature of the lower air layers increases considerably almost over the entire surface, due to the greenhouse effect caused by the absorption of long-wave radiation by water vapor and carbon dioxide in the atmosphere. This effect is not significant at very low temperatures, and the formation of dense clouds that change the radiative flux is also impossible. Under these conditions, the atmosphere becomes more or less transparent to both shortwave and long-wave radiation.

The mean temperature of the earth's surface when the atmosphere is transparent is determined by the simple formula $[S_0(1 - \alpha_s)/4\sigma]^{1/4}$, where S_0 is the solar constant and σ is the Stefan-Boltzman constant. It follows from this formula that when $\alpha_s = 0.80$, the earth's mean temperature is 186°K, or −87°C.

Thus, if the entire earth's surface were covered with snow and ice, even for a short period, its mean temperature, which is now equal to 15°C would drop by approximately 100°C. This example illustrates what a tremendous influence the snow cover has on the thermal regime.

We have made an attempt to evaluate the effect of polar sea ice on the thermal regime in the Arctic (Budyko, 1961, 1962, 1971). By using the data on the heat balance of central regions of the Arctic Ocean and approximate formulas of the semiempirical theory of climate, it was established that the polar ice cools the air in the Central Arctic by several degrees in the summer and by 20°C in the winter.

We have come to the conclusion that at present an ice-free regime could exist in the Arctic Ocean, although it would be very unstable, and minor climatic fluctuations could lead to the restoration of the ice cover. Rakipova (1962, 1966), Donn and Shaw (1966) and Fletcher (1966) also came to the conclusion that an ice-free regime in the Arctic was possible.

Thus, a permanent ice cover occupying even a comparatively small fraction of the earth's surface has a strong influence on the atmosphere's thermal regime, which must be taken into account in the study of climatic variations.

Semiempirical Model of the Thermal Regime. In spite of all the outstanding achievements in the development of a general theory of climate, it is still difficult to apply existing models to the study of climate changes. This is due to the fact that these models do not, as a rule, meet all the abovementioned requirements. At the same time, as pointed out by Smagorinsky (1974), computations of climatic variations using models of the general theory of climate require extensive computer time and are therefore usually far beyond the capabilities of computer centers. In this connection Smagorinsky noted the necessity of a parametric representation of the circulation processes in the atmosphere and ocean in order to statistically describe large-scale perturbations, as is done in the general study of small-scale turbulent processes. This idea has been applied in several works in which heat and humidity transfer processes in the atmosphere were treated as large-scale turbulent processes.

In view of the difficulties inherent in applying the general theory of climate to the study of its variations, several attempts have been made to use semiempirical models. These models can be made to meet the requirements by means of extensive parameterization of large-scale atmospheric processes, by limiting the given problems, and by using empirical relations.

This trend includes a semiempirical model of the thermal regime of the atmosphere (Budyko, 1968; Budyko and Vasishcheva, 1971), based on a number of considerations which we shall discuss below.

In the absence of an atmosphere, the mean temperature of the earth's surface would be defined by the conditions of radiation equilibrium, i.e. the long-wave radiation of the earth's surface is equal to the absorbed radiation. Let us express this condition in the following form:

$$\delta \sigma T^4 = \tfrac{1}{4} S_0 (1 - \alpha_s), \tag{2.4}$$

where δ is a coefficient that characterizes the properties of the radiation surface as distinct from that of a black body, σ the Stefan constant, T the surface temperature, S_0 the solar constant, and α_s the mean albedo of the earth.

Assuming that $\delta = 0.95$, $\sigma = 8.14 \cdot 10^{-11}$ $cal/cm^2 \cdot min$, and $S_0 = 1.95$ $cal/cm^2 \cdot min$, it follows from (2.4) that for $\alpha_s = 0.33$, the mean earth temperature is 255°K, or -18°C. Since the mean air temperature near the earth's surface, according to observations, is about 15°C, we arrive at the conclusion that with the same albedo, the atmosphere increases the mean air temperature at the earth's surface by approximately 33°C. This temperature increase is caused by the greenhouse effect, i.e. by the higher transparency of the atmosphere to the short-wave, as compared to the long-wave radiation.

It must be noted that this evaluation is true under certain conditions, namely, in the absence of atmosphere the planetary albedo cannot be equal to the actual value of 0.33.

Under present conditions, the mean albedo of the earth's surface is 0.14. It may be assumed that prior to the origin of the atmosphere, the earth's albedo was lower and probably differed very little from the moon's albedo, which is equal to 0.07. With this albedo the mean temperature of the earth's surface could be about 3°C.

In order to estimate the influence of incoming solar radiation and the albedo on the mean temperature at the earth's surface under existing conditions, it is necessary to know the dependence on temperature distribution of long-wave radiation at the outer boundary of the atmosphere. This dependence can be established in two ways.

The first involves considering a theoretical model of the vertical distribution of radiation fluxes, temperature and air humidity. The second is based on an empirical comparison of data on long-wave radiation obtained from observations or calculations, and the factors that affect it.

This method has been applied by the author (Budyko, 1968), who utilized the average monthly values of the outgoing radiation calculated by Vinnikov in the course of compiling maps for the Atlas of Heat Balance of the World (1963).

These included data on outgoing radiation obtained at 260 stations uniformly distributed on land and sea. We had at our disposal 3120 average monthly values.

A comparison of these data with various elements of the meteorological regime revealed that these average monthly values depended mainly on the air temperature near the earth's surface and on cloudiness. This relation was expressed in the following form:

$$I_s = a + bT - (a_1 + b_1 T)n, \qquad (2.5)$$

where I_s is the outgoing radiation in kcal/cm$^2 \cdot$ month; T is air temperature in °C; n is cloudiness in fractions of a unit, and the atmospheric dimensional coefficients are: $a = 14.0$; $b = 0.14$; $a_1 = 3.0$; $b_1 = 0.10$.

The mean deviation of the outgoing radiation for each month, from the results obtained by (2.5), was very small (below 5%), which confirms the conclusion that all other factors, including the peculiarities of the vertical temperature distribution, exercise a comparatively small influence on the outgoing radiation.

It must be pointed out that this deduction applies to values of outgoing radiation averaged over long periods, and that it refers to large geographical areas. The physical interpretation of the above conclusion is that the outgoing radiation originates mainly in the troposphere where the temperature deviations from its mean vertical distribution are generally lower than its annual spatial variations.

Equation (2.5) can be compared to the results of several theoretical works including, in particular, that of Manabe and Wetherald (1967). In this work a relation was established between the outgoing radiation and temperature and cloudiness under radiational and convectional equilibrium conditions in the atmosphere. This dependence was presented graphically, its curves could be expressed by the formula:

2 THE GENESIS OF THE CLIMATE

$$I_s = 14 + 0.14T - 1.6n. \tag{2.6}$$

We can see that the independently obtained formulas (2.5) and (2.6) coincide in the case of a clear sky.

To verify the formulas showing the connection between the outgoing radiation and meteorological factors, we must apply the equilibrium condition between the outgoing and absorbed radiation for the entire planet:

$$Q_{sp}(1 - \alpha_{sp}) = I_{sp}, \tag{2.7}$$

where the values Q_{sp}, α_{sp}, and I_{sp} apply to the planet as a whole.

Assuming that the average for the entire earth value of Q_{sp} is equal to 20.8 $kcal/cm^2 \cdot month$, the average temperature at the earth's surface level is 15°C and the mean cloudiness is 0.50, we find from formula (2.5), that condition (2.7) holds when $\alpha_{sp} = 0.33$. This value of the planet's albedo agrees with the estimates obtained by independent methods in several modern investigations, thus confirming the accuracy of formula (2.5).

It can be shown, however, that in accordance with the above statements concerning the effect of snow and ice covers on the thermal regime, equations (2.5) and (2.7) have at least one more solution.

Let us assume that at low negative air temperatures at the earth's surface at all latitudes, it is entirely covered with snow and ice, and that the global albedo of the earth-atmosphere system in this case is approximately equal to that of the central Antarctic region, i.e. 0.6–0.7. From (2.5) and (2.7) it follows that in this case the average planetary temperature at the earth's surface will vary from $-47°C$ (at the lower albedo value) to $-70°C$ (at the higher albedo value). At these average temperatures the total glaciation of the earth is inevitable, thus confirming the above remark as to the ambiguous relation between contemporary climatic conditions and external climatic factors. This problem will be dealt with in greater detail further on.

Let us find from (2.5) and (2.7) the expression for the average earth temperature:

$$T_p = \frac{1}{b - b_1 n} [Q_{sp}(1 - \alpha_{sp}) - a + a_1 n]. \tag{2.8}$$

This formula is correct within the limits of the actual fluctuations of mean monthly temperatures at the earth's surface, its precision depending on that of expression (2.5). We must bear in mind that in accordance with the structure of formula (2.8), the precision of temperature calculations for heavy cloudiness (i.e. for n close to unity) is considerably lower than for low or medium cloudiness. Although this limits the possibilities of applying formula (2.8), certain conclusions may be drawn as to the effect of cloudiness on the thermal regime of the lower atmospheric levels.

For this purpose we must take into consideration the dependence of albedo on cloudiness, which has the form

$$\alpha_s = \alpha_{s_n} n + \alpha_{s_0}(1 - n), \qquad (2.9)$$

where α_{s_n} and α_{s_0} are the albedo values of the earth-atmosphere system for overcast cloudiness and a clear sky, respectively.

From (2.8) and (2.9) it follows that

$$T_p = \frac{1}{b - b_1 n} \{Q_{sp}[1 - \alpha_{s_n p} n - \alpha_{s_0 p}(1 - n)] - a + a_1 n\}. \qquad (2.10)$$

Assuming, in accordance with the method applied in compiling the maps for the Atlas of Heat Balance of the World (1963), that

$$\alpha_{s_0 p} = 0.66\alpha + 0.10 \qquad (2.11)$$

where α is the albedo of the earth's surface, we find for the northern hemisphere an average value of $\alpha_{s_0 p} = 0.20$. Then from (2.9) we infer that for $\alpha_{sp} = 0.33$ and $n = 0.50$, the average value $\alpha_{s_n p} = 0.46$.

Taking into account the obtained values of $\alpha_{s_0 p}$ and $\alpha_{s_n p}$, from (2.10) we find that for an average global value of Q_{sp} the effect of cloudiness on temperature is comparatively small and probably lies within the limits of precision, which is not high in this case.

A more detailed analysis of the relation between cloudiness and the air temperature at the earth's surface is given by Schneider (1972), from which it follows that this influence strongly depends on the altitude of the upper boundary of the clouds. Since the relation between the clouds' altitude and the temperature has not been sufficiently studied, it was impossible to determine the sign of this relation under average global conditions. Subsequently the question of the effect of cloudiness changes on the mean air surface temperature was studied in the author's works (Budyko, 1975) and of Cess (1976), based on the analysis of satellite data on outgoing long-wave radiation. The conclusion was derived in these works that this effect is rather small.

Formula (2.10) makes it possible to draw certain conclusions as to the effect of climatic factors on the average temperature at the earth's surface.

A 1% variation in the solar radiation changes the average temperature by approximately 1.5°C for cloudiness equal to 0.50 and the actual albedo value. This estimate is about twice as high as that of temperature variations in the absence of atmosphere. Thus the radiational properties of the atmosphere increase the effect of radiational fluctuations on the termperature distribution of the earth's surface.

The variation of albedo by 0.01 changes the temperature by 2.3°C. Hence the thermal regime greatly depends on the albedo if its variation is not caused by cloudiness. As stated above, the effect of cloudiness on the thermal regime connected with albedo variations is compensated to a great extent by corresponding variations of the outgoing long-wave radiation.

Model of an Average Latitudinal Temperature Distribution. Horizontal redistribution of heat in the atmosphere and hydrosphere strongly influences the thermal regime at certain latitudinal zones.

This problem is very difficult to deal with since all the forms of horizontal heat transfer in the atmosphere and hydrosphere which are comparable with the amount of absorbed solar radiation have to be taken into consideration. They include heat transfer by mean and large-scale turbulent motions in the atmosphere and hydrosphere as well as heat redistribution due to phase transformations of water. Since existing theoretical models are not yet sufficiently developed, it is difficult to determine precisely the parameters of horizontal heat distribution, in order to take into account all these forms of heat transfer. Data on the heat balance components may be utilized in the solution of this problem.

The heat balance equation of the earth-atmosphere system has the form

$$Q_s(1 - \alpha_s) - I_s = C + B_s, \qquad (2.12)$$

where $C = F_s + L(E - r)$, i.e. the sum of heat inflow due to horizontal motions in the atmosphere and hydrosphere.

For annual mean conditions, the term B_s corresponding to the gain or loss of heat during a given period is equal to zero, and the heat influx C is equal to the radiation balance of the earth-atmosphere system. Since the values of this balance can be determined either by observations or calculations, we can simultaneously determine the value of the horizontal heat redistribution.

It may be assumed that the value of C is in some way connected with the horizontal distribution of the mean temperature in the troposphere. Considering that the deviation of the earth's temperature from the mean vertical distribution in the troposphere is small compared to geographic temperature fluctuations, we find that the mean air temperature in the troposphere is closely connected with the temperature at the earth's surface. This assumption is fully confirmed by a comparison of average monthly temperatures at the earth's surface with those at the altitude of the 500 mb level for various geographical regions and seasons (Kagan and Vinnikov, 1970).

Thus there exists a dependence between horizontal heat transfer and temperature distribution near the earth's surface.

Let us analyze the relation between these parameters for latitudinal zones of the northern hemisphere under mean annual conditions. For this purpose let us calculate the radiation balance using the equation

$$R_s = Q_s(1 - \alpha_s) - I_s. \tag{2.13}$$

For latitudes 0–60° we shall determine the albedo according to formulas (2.9) and (2.11), while for polar latitude, where formula (2.11) is not sufficiently precise, we shall use albedo values obtained from satellite observations (Raschke, Möller and Bandeen, 1968).

It is possible to determine the meridional heat flux in the earth-atmosphere system from the values thus obtained for the radiation balance of the earth-atmosphere system equal to the term C. A comparison of these values to relevant air temperature gradient at the earth's surface shows that the relation between the parameters differs at various latitudes. Thus using the assumption that the meridional heat flow is proportional to the average latitudinal temperature gradient, we must assume that the proportionality factor, i.e. the coefficient of large-scale turbulent heat exchange, is a function of latitude. Although, under present conditions, this function can be determined empirically, its behavior under changing climatic conditions remains unknown. Therefore it is necessary to find other empirical relations between the parameter C and the temperature distribution.

Considering that meridional heat transfer is directed from warmer to colder regions, we may say that C depends on $T - T_p$, where T is the mean temperature at a given latitude and T_p is the mean global temperature of the lower atmospheric level.

A similar assumption was used by Öpik (1953) and Sawyer (1963, 1966).

To investigate this dependence let us compare the obtained values of R_s with corresponding values of $T - T_p$ (see Fig. 5). We find that there is a definite relation between the parameters that can be expressed by the empirical formula

$$Q_s(1 - \alpha_s) - I_s = \beta(T - T_p), \tag{2.14}$$

where $\beta = 0.235 \ kcal/cm^2 \cdot month \cdot deg$.

This dependence greatly simplifies the estimate of the meridional heat redistribution in the model of the thermal regime.

From (2.5) and (2.14) we find that

$$T = \frac{Q_s(1 - \alpha s) - a + a_1 n + \beta T_p}{\beta + b - b_1 n}, \tag{2.15}$$

from which it is possible to calculate the mean temperature at various latitudes, these results being presented in Fig. 6 by the T_0 curve

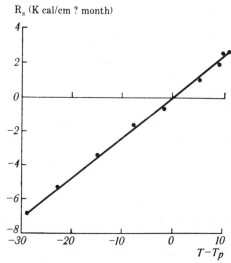

Fig. 5. Dependence of meridional heat transfer on temperature difference.

together with the T curve for measured temperatures. Obviously the calculated and observed temperatures correlate well.

It should be borne in mind that this correlation was obtained by using a model containing only a single empirical parameter (β) which depends on the temperature distribution but not on the latitude.

Note that in the case of a polar ice cover boundary concentric to the

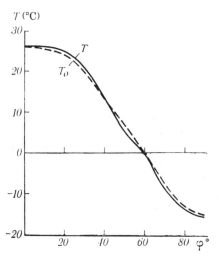

Fig. 6. Mean latitudinal distribution of air temperature; T_0, calculated values; T, measured values.

pole, we obtain spasmodic variations of the zonal-mean air temperature at the ice field boundary; these variations correspond to variations in the albedo of the earth-atmosphere system. But since the actual ice boundary is not concentric, it is advisable to calculate the average latitudinal air temperature by interpolating it in the zone from the lowest to the highest latitude of the ice boundary thus eliminating its spasmodic variation within this latitudinal belt.

Observational data confirm that the temperature jump on the ice boundary, implied by calculations according to this model, actually take place in various sectors of the Arctic seas during the cold season. It may be assumed that in the case of a concentric ice distribution, this temperature jump would have been pronounced in the distribution of winter temperatures derived from observational data.

Model for Various Seasons. In order to study the atmosphere's thermal regime during various seasons, it is necessary to modify this model by taking several additional factors into account.

Such a more general model was suggested by Budyko and Vasishcheva (1971) with the heat balance equation of the earth-atmosphere system having the form

$$Q_{sw}(1 - \alpha_{sw}) - I_{sw} = C_w + B_s, \tag{2.16}$$
$$Q_{sc}(1 - \alpha_{sc}) - I_{sc} = C_c - B_s, \tag{2.17}$$

where B_s is the heat gain or loss due to cooling or heating of the earth-atmosphere system which depends mainly on the process of cooling or heating of the ocean. Here and further on, the subscripts "w" and "c" correspond to the warm and cold half-years.

We shall make use of equations (2.5), (2.12), and (2.14) in order to determine the I_s and C terms.

The use of the last equation is made possible by a comparison of the values of meridional heat redistribution for various seasons, determined from data on the heat balance of the earth-atmosphere system, using the appropriate air temperature differences at the earth's surface. As a result of this comparison, it was established that relation (2.14) holds for various seasons, the β proportionality factor increasing slightly for the cold, and dropping for the warm half-year. The value of this factor for the warm period in the southern hemisphere equaled that for the northern hemisphere, while for the cold period it was higher than that for the northern hemisphere.

The value of B_s is found from

$$B_s = s\gamma(T_{ww} - T_{wc}), \tag{2.18}$$

where T_{ww} and T_{wc} are average latitudinal temperatures of the ocean surface waters for the warm and cold half-years, respectively; s is a fraction of the total area of a given latitudinal zone occupied by the ocean, and γ a dimensional factor.

In order to determine the ocean surface temperature we use its heat balance equation in the form

$$R_{ww} = LE_w + P_w + \frac{B_s}{s}, \qquad (2.19)$$

$$R_{wc} = LE_c + P_c - \frac{B_s}{s} + F_0, \qquad (2.20)$$

where R_w is the radiation balance of the ocean surface; LE the heat expenditure for evaporation; P the turbulent heat flux between the ocean surface and the atmosphere; F_0 is the heat transfer by ocean currents which, according to empirical data, is of the utmost importance during the cold time of the year, when its absolute value is comparable to that of the radiation balance.

To determine LE, P and F_0, we apply the approximate relations

$$LE = fT_w, \qquad (2.21)$$
$$P = c(T_w - T), \qquad (2.22)$$
$$F_0 = \beta'(T_c - T_p), \qquad (2.23)$$

where T_w is the surface temperature of ocean water in °C, and T_c is air temperature for the cold half-year, f, c and β' dimensionless factors.

The approximate relations (2.18), (2.21) and (2.22) were obtained through simplifying the equations describing the heat exchange processes in the upper ocean levels and on its surface. Formula (2.23) was obtained by a method similar to the one used for deriving the formula $C = \beta(T - T_p)$.

In obtaining expression (2.21) we have simplified the well-known evaporation formula

$$E = A(V)(q_w - q),$$

where q_w is the specific air humidity equal to its saturation value at the temperature of the ocean surface, q is the specific air humidity, and $A(V)$ is a coefficient depending on wind velocity. If we neglect the dependence of coefficient A on the wind velocity V and the deviations of q/q_w from the mean value, then, by using the relation between the saturation humidity values and temperatures, we arrive at the approximate relation (2.21).

Our calculations involve only average latitudinal and average seasonal values of the pertinent terms of the heat balance, and, as can be seen from numerical experiments, the results of calculations of temperature distribution depend very little on the errors of the heat balance components used.

This makes it possible to neglect the influence on the evaporation from the ocean surface, of changes in the relative humidity over the ocean and in the wind velocity in various latitude zones, thus simplifying the determination of heat expenditure for evaporation and calculations of air temperature, without substantially affecting the precision of obtained results.

The numerical factors in these equations (when the heat flows are determined in kcal/(cm² month) have the following values: $\gamma = 3.0$; $f = 0.4$; $\beta_w = 0.22$; $\beta_c = 0.27$ for the northern hemisphere south of the mean boundary of Arctic ice; $\beta_c = 0.40$ for the southern hemisphere north of the mean boundary of Antarctic ice; $\beta_c = 0.22$ for zones with an ice cover; $\beta' = 0.14$ for the northern hemisphere; $\beta' = 0.20$ for the southern hemisphere; $c = 0.84$.

From this relation we may obtain the following formulas for determining the average latitudinal temperatures in the northern hemisphere:

$$T_w = \frac{Q_{sw}(1-\alpha_{sw}) - a + a_1 n_w + \beta_w T_{p1} - \dfrac{\gamma s}{f+c+2\gamma} \cdot (R_{ww} - R_{wc} - \beta' T_{p2}) - \dfrac{\gamma s(\beta'-c)}{f+c+2\gamma} Q_{sw}(1-\alpha_{sw}) + Q_{sc}(1-\alpha_{sc}) \dfrac{-2a + a_1 n_w + a_1 n_c + \beta_w T_{p1} + \beta_c T_{p2}}{b - b_1 n_c + \beta_c}}{b - b_1 n_w + \beta_w + \dfrac{\gamma sc}{f+c+2\gamma} - \dfrac{\gamma s(\beta'-c)(b - b_1 n_w + \beta_w)}{(f+c+2\gamma)(b - b_1 n_c + \beta_c)}} \quad (2.24)$$

$$T_c = \frac{Q_{sc}(1-\alpha_{sc}) - a + a_1 n_c + \beta_c T_{p2} + \dfrac{\gamma s}{f+c+2\gamma} \cdot (R_{ww} - R_{wc} + \beta' T_{p2}) + \dfrac{\gamma sc}{f+c+2\gamma} Q_{sw}(1-\alpha_{sw}) + Q_{sc}(1-\alpha_{sc}) \dfrac{-2a + a_1 n_w + a_1 n_c + \beta_w T_{p1} + \beta_c T_{p2}}{b - b_1 n_w + \beta_w}}{b - b_1 n_c + \beta_c + \dfrac{\gamma s(c-\beta') + \gamma sc(b - b_1 n_c + \beta_c)}{f+c+2\gamma(f+c+2\gamma)(b - b_1 n_w + \beta_w)}} \quad (2.25)$$

In this and the following formulas subscript "p" of values T, Q_s, n, α_s, R, s refers to the planet as a whole, subscript "1" to the warm half-

2 THE GENESIS OF THE CLIMATE

year in the northern hemisphere and to the cold half-year in the southern hemisphere, and subscript "2" to the cold half-year in the northern hemisphere and the warm half-year in the southern hemisphere.

Thus formulas (2.24) and (2.25) can also be used for calculating average latitudinal temperatures in the southern hemisphere as well, assuming in this case that T_{p1} refers to the cold half-year in the southern hemisphere and T_{p2} to the warm half-year in the same hemisphere.

For an ice-covered zone formulas (2.24) and (2.25) are somewhat simplified. Since in this zone the annual change of heat content in the earth-atmosphere system is small and the meridional heat transfer by currents is negligible, if it exists at all we find that we may use the equations

$$T_w = \frac{Q_{sw}(1 - \alpha_{sw}) - a + a_1 n_w + \beta_w T_{p1} - lh}{b - b_1 n_w + \beta_w}, \qquad (2.26)$$

$$T_c = \frac{Q_{sc}(1 - \alpha_{sc}) - a + a_1 n_c + \beta_c T_{p2} + lh}{b - b_1 n_s + \beta_c}, \qquad (2.27)$$

where lh is the gain (or loss) of heat due to cooling (or warming) of the ice cover and the accumulation (or melting) of ice.

Since the value of lh is comparatively small, we shall consider its average value for all latitudes and zones as equal to 0.8 $kcal/cm^2 \cdot month$.

To determine average planetary temperatures we shall use formulas which can be derived from (2.24) and (2.25), assuming that for the planet as a whole the meridional heat exchange is equal to zero;

$$T_{p1} = \frac{\begin{array}{c} Q_{sp1}(1 - \alpha_{sp1}) - a + a_1 n_{p1} - \dfrac{\gamma s_p}{f + c + 2\gamma} \\[6pt] \cdot (R_{wp1} - R_{wp2}) + \dfrac{\gamma s_p c}{f + c + 2\gamma} \\[6pt] \dfrac{Q_{sp1}(1 - \alpha_{sp1}) + Q_{sp2}(1 - \alpha_{sp2}) - 2a + a_1 n_{p1} + a_1 n_{p2}}{b - b_1 n_{p2}} \end{array}}{b - b_1 n_{p1} + \dfrac{\gamma s_p c}{f + c + 2\gamma} + \dfrac{\gamma s_p c(b - b_1 n_{p1})}{(f + c + 2\gamma)(b - b_1 n_{p2})}}, \qquad (2.28)$$

$$Q_{sp2}(1 - \alpha_{sp2}) - a + a_1 n_{p2} + \frac{\gamma s_p}{f + c + 2\gamma}$$

$$\cdot (R_{wp1} - R_{wp2}) + \frac{\gamma s_p c}{f + c + 2}$$

$$T_{p2} = \frac{Q_{sp1}(1 - \alpha_{sp1})(1 - \alpha_{sp2}) - 2a + a_1 n_{p1} + a_1 n_{p2}}{b - b_1 n_{p1}} \qquad (2.29)$$
$$b - b_1 n_{p2} + \frac{\gamma s_p c}{f + c + 2\gamma} + \frac{\gamma s_p c(b - b_1 n_{p2})}{(f + c + 2\gamma)(b - b_1 n_{p1})}$$

By applying formulas (2.24)–(2.29), we can calculate the distribution of average latitudinal temperatures for every half-year.

In this calculation it was assumed that the boundary of polar ice corresponds to the average latitudinal temperature of the warm half-year, which is equal to −1°C. For the polar-ice zone the albedo at the earth-atmosphere system was taken to be 0.62 and 0.72 for the northern and southern hemispheres respectively. The albedo of latitude zones without any permanent snow or ice cover was determined according to Table 8. The values given in this table were obtained from satellite observations (Raschke, Möller, and Bandeen, 1968).

In determining the total radiation incident on the outer boundary of the troposphere, the radiation flux incident on a unit surface perpendicular to the solar beam (meteorological solar constant) was assumed to be 1.92 $cal/cm^2 \cdot min$. The average latitudinal values of the radiation balance for the ocean surface were taken from the Atlas of Heat Balance of the World (1963).

These results are presented in Fig. 7 together with the distribution of average latitudinal temperatures obtained by observations. As we see, the discrepancy between the measured and calculated temperatures at various latitudes is at most 1–2°.

The described numerical model of the atmospheric thermal regime for various seasons is not the only one. In a number of studies several other numerical models have been suggested, in many cases more detailed than this one. These models, however, did not as a rule meet all the requirements listed at the beginning of this section.

Table 8
Albedo of Latitude Zones

Latitude (°N)	Half-Year		Latitude (°S)	Half-Year	
	First	Second		First	Second
65	0.49	0.54	5	0.25	0.27
55	0.40	0.48	15	0.25	0.26
45	0.35	0.45	25	0.29	0.28
35	0.29	0.40	35	0.40	0.32
25	0.27	0.30	45	0.46	0.41
15	0.27	0.25	55	0.53	0.49
5	0.29	0.26			

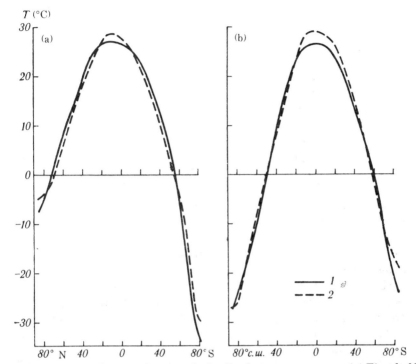

Fig. 7. Distribution of average latitudinal air temperatures. (a) First half-year; (b) second half-year. Solid line, observational data; dashed line, calculated data.

Our model of the thermal regime satisfies all these requirements to a first approximation, so that it is possible to use it in calculating climatic changes.

Unambiguity and Stability of Climate. We may analyze the unambiguity and stability of climate by using these models of the atmospheric thermal regime.

Let us assume, in accordance with empirical data obtained both for polar sea ice and for continental glaciations on a plain and in mountains outside of the tropical zone, that the mean latitudinal boundary of the permanent snow and ice cover corresponds to an average annual temperature of $-10°C$. Under this condition we will calculate (by means of the thermal regime model for average annual conditions) the dependence of the mean boundary of the polar ice cover in the northern hemisphere as a function of the solar radiation at the upper boundary of the atmosphere.

This dependence is depicted in Fig. 8, where $\Delta Q_{sp}/Q_{sp}$ is the relative variation of the solar radiation at the outer boundary of the atmosphere,

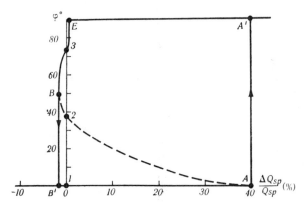

Fig. 8. Dependence of the mean latitude of the polar-ice boundary in the northern hemisphere on the radiation influx.

in per cent, and φ is the mean latitude of the polar ice boundary in the northern hemisphere.

The dependence of φ on $\Delta Q_{sp}/Q_{sp}$ is presented by a set of heavy lines showing that this relation is ambiguous and that it differs substantially in the cases of increasing and decreasing influx of solar radiation.

First let us consider the case when heat influx increases from an initial low value. When the increase is small, we have complete glaciation of the earth ($\varphi = 0°$), which is maintained until the heat influx reaches its present value (point 1) and up to a value increasing the given value by several tens of per cent (point A). The glaciation regime corresponding to point A is unstable and when the heat influx slightly increases, it changes to a regime where glaciation is completely absent (point A'). The ice-free regime is preserved with a further increase of the heat influx.

If the heat influx decreases below the initial value which greatly exceeds the solar constant, then we have, at first, conditions for an ice-free regime ($\varphi = 90°$). When the curve reaches point E, which is near the current value of the heat influx, polar glaciation occurs, which grows rapidly as the heat influx decreases. After reaching point 3, corresponding to the current climatic regime, the decrease of the heat influx by about 2% causes the ice cover to spread southward to latitude 50°N (point B). The glaciation regime corresponding to this point is unstable, and with a small decrease of heat influx, changes to complete glaciation (point B'), which remains as the heat influx further decreases.

By means of this model we can also obtain the relation between φ and $\Delta Q_{sp}/Q_{sp}$ depicted by the dashed line AB. Since the curve corresponds to growth of φ as $\Delta Q_{sp}/Q_{sp}$ decreases, and vice versa, it also corresponds to unstable glaciation regimes, which are replaced by a total glaciation regime or an ice-free regime, with small fluctuations of heat influx. Consequently, point 2 on curve AB representing the actual value of the heat influx, indicates an unstable regime that cannot exist for an extended period of time.

Thus the relation between the polar glaciation and the heat influx bears the form of a hysteresis loop, whose sections AA' and BB' correspond to the transition from one solution of the given equations to the other, as shown in Fig. 8 by arrows. Other sections of the hysteresis loop corresponding to stable regimes represent the dependence of φ on $\Delta Q_{sp}/Q_{sp}$, both for increase and decrease of the heat influx.

In studying the relation given in Fig. 8, we can also use the thermal regime model for various seasons. The use of this model yields results that quantitatively differ somewhat from the diagram in Fig. 8 but entirely confirm the main qualitative regularities. This could be expected, since these regularities can be established on the basis of general considerations that must be taken into account in compiling various models of the thermal regime.

As shown above, the present value of the solar constant does not exclude the possibility of a stable and complete glaciation of the planet with very low temperatures at all latitudes (white earth). The stability of this regime is due to the very high albedo of a surface covered with snow and ice. In this case climatic conditions typical for the Antarctic might prevail over the entire surface.

The possibility of a stable white earth regime apparently follows from any realistic theory of climate.

The mean air temperature at the surface of a white earth can vary greatly, depending on the albedo of the earth's atmosphere system over a planet covered with snow and ice. Since the albedo can be calculated only approximately, existing estimates of the mean temperature vary widely. However, in all cases it appears to be much below the water-freezing point, which indicates a high stability for a thermal regime under these conditions.

Thus the current influx of solar radiation permits a complete glaciation of the earth; this regime is designated in Fig. 8 by point 1. It is also possible if the level of incoming radiation drops below the current level (as shown in Fig. 8 by a line running along the horizontal axis to the left of point 1), and also if the incoming radiation increases above the present level up to a value corresponding to the ice-melting temperature in the warmest regions of the earth's surface. Once this temperature is reached, part of the surface may get clear of ice, reducing the albedo and increasing the absorbed radiation. The point corresponding to the boundary of the complete glaciation regime with increasing solar radiation is indicated in Fig. 8 by A.

It is natural to assume that, between the conditions outlined at points A and 3, there exist regimes of partial glaciation which can be shown on this diagram by a line connecting these points. The path of this line may be determined in the following manner.

Observational data as well as general physical considerations suggest that under present climatic conditions, an increase of the heat influx to the earth's surface would lead to a rise in the mean temperature at the surface and to a recession of polar ice, while a decrease of the heat influx would cause a drop in the mean temperature and an expansion of the ice sheet.

In accordance with this rule, the line running from A to 3 must approach this point from the left, which is possible if this line intersects the vertical axis at least at one more point (point 2).

Thus we arrive at the conclusion that a third climatic regime may exist under present conditions, that is a second version of the partial glaciation regime of the earth, the surface of the ice sheet being larger than one we have now.

After intersecting the vertical axis at point 3, the line corresponding to the regime of partial glaciation, at a certain value of incoming radiation exceeding its present amount, will reach the horizontal line corresponding to the ice-free regime (point E). Further on, the line corresponding to a continuing radiation increase, becomes horizontal and exits at the right.

In several of the mentioned works, calculations have been made proving that with the current radiation influx, an ice-free regime is probable in the Arctic. If we agree with this contention, then this regime can be represented on our diagram by a point on the vertical axis at $\varphi = 90°$, and in this case the line for the partial glaciation regime will approach the horizontal line corresponding to the ice-free regime, not from the right of this point, but from the left. This position of the intersection is caused by the fact that if an ice-free regime is possible under present conditions, it is also possible under slightly lower incoming radiation.

Thus, in this case the line representing a possible glaciation regime passes through the given point as a horizontal line that must cross the vertical axis at least once more. Therefore, with conditions when an ice-free regime can exist in the Arctic at the present level of incoming solar radiation, there must exist another regime of partial glaciation with a smaller surface of ice cover than is the case for present climatic conditions.

In addition to the relation presented in Fig. 8, we show in Fig. 9 a relation between the average planetary air temperature and the heat

2 THE GENESIS OF THE CLIMATE

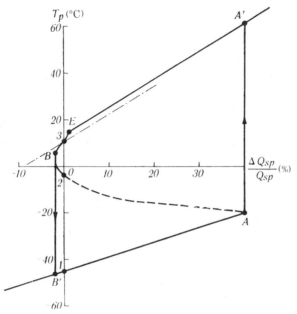

Fig. 9. Dependence of the average planetary air temperature on the radiation influx.

inflow, calculated according to the same model. All designations in this figure are the same as in Fig. 8.

It follows from Fig. 9 that this relation has much in common with that for the boundary of polar ice, and also has the form of a hysteresis loop. The result presented in Figs. 8 and 9 strongly depends on the albedo difference of the earth-atmosphere system with and without a snow and ice cover. The albedo values used in these calculations were determined from satellite observations. In this connection it should be noted that the new reduction of these observations (Raschke, et. al., 1973) produced somewhat lower albedo values for the polar regions than the earlier works which were used in our calculations. The estimate, based on the new data, somewhat modified the quantitative results presented in Figs. 8 and 9, but did not affect their main character.

Our conclusion that modern climate is ambiguous is based on the use of a thermal regime model for the stationary condition of the atmosphere-ocean-ice cover system. This condition corresponds to a stable or slowly changing climate.

For a climate whose changes are comparatively rapid, it may not be appropriate to assume that the system is stationary. Since the continental ice covers as compared to other parts of the system are markedly stable, a greater number of different climatic versions can exist under nonstationary conditions as compared to the few solutions corresponding to stationary conditions.

Let us assume that during the postglacial warming there still existed a continental glaciation with a boundary not corresponding to the changed climatic conditions. In this case a new type of thermal regime appears, which can be calculated by means of a numerical model for a given ice-cover boundary. This thermal regime changes together with the ice-cover boundary, producing an unlimited number of climatic versions corresponding to the given values of the external climatic factors.

However, despite the fact that in our epoch there exist such vast glaciations as the Arctic and Antarctic, the polar ice can be approximately considered stationary, since their mean boundaries in the northern and southern hemispheres correspond to the same values of the climatic factors that determine these boundaries.

The stability of the current mean boundary of the polar ice in both hemispheres is apparently caused by the fact that the main part of this boundary consists of not very thick sea ice.

The actual temperature variations during the main part of the earth's history were limited to short sections of the lines in Figs. 8 and 9, from just above point E to just above point B.

It is interesting to compare this part of the line in Fig. 9 to the dot-and-dash line passing through point 3 and corresponding to variations of the average planetary temperature, when the polar ice does not affect the thermal regime. This comparison shows to what extent the polar ice intensifies the temperature variations caused by fluctuations in incoming heat.

Thus the present climate apparently is not the only possible one under the existing external climatic factors, i.e. it is not transitive in the sense of the definition given by Lorentz. It is possible that in addition to the considered mechanism there are also other forms of climate intransitivity that will be found in further studies.

In order to estimate the stability of the current climate, we may make use of the relation presented in Figs. 8 and 9. As we see, the climatic regimes corresponding to sections BE are highly sensitive to small variations of the heat influx. It follows therefore that comparatively small variations in external climatic factors can considerably affect the climate of today.

This conclusion is of great importance for understanding the laws of climatic changes.

Results of Other Investigations. After the first work on the stability of current climate had been published (Budyko, 1968) this problem

2 THE GENESIS OF THE CLIMATE

was studied in several other investigations, the results of which we shall now analyze. The first of these studies was carried out independently of our work by Sellers (1969). Sellers compiled a semiempirical model of the atmosphere's thermal regime which was based on the energy balance equation of the earth-atmosphere system and took into consideration the feedback between the thermal regime and polar ice. From this model Sellers determined the dependence of the average latitudinal temperature at the earth's surface upon the solar constant, the albedo of the earth-atmosphere system, and other climatic factors. Then he used these relations to estimate the climatic effects of the polar ice and solar radiation.

From Sellers' work it followed that if the albedo of the earth-atmosphere system in the Central Arctic drops from 0.62 to 0.32 as a result of melting polar ice, the average annual air temperature at the earth's surface increases in this region by 13°C, while at low latitudes it increases by 2°C. In calculating the decrease in the solar constant sufficient to cause the complete glaciation of the earth, Sellers found, on the basis of the albedo-temperature relations obtained from observational data, that it amounted to 2%. For a weaker albedo-temperature dependence, this value increased to 5%. However, Sellers considered that this relation did not agree with observations.

It is noteworthy that Sellers' results are closely correlated with those of our paper, which likewise state that the existing ice cover reduces the average annual temperature in the Central Arctic by 14°C and at low latitudes by 2°C.

This correlation of the main results is very interesting, because, although the main principle of the climate models was the same, the calculations were based on many different assumptions (Budyko, 1970; Sellers, 1970).

In his next work, Sellers (1973) developed a more detailed model of the atmosphere's thermal regime for different seasons, the main results of which were close to those of his previous works.

Another semiempirical model of the atmospheric thermal regime that takes into account the feedback between the temperature and the polar ice was developed by Faegre (1972). Faegre confirmed the main deductions of the above works and called attention to the fact that his model indicated the ambiguity of the current climate. Faegre obtained five different solutions of the equation for the average latitudinal temperature distribution at the earth's surface. Out of these he selected three as having physical sense: 1) for existing climatic conditions, 2) for the white earth, and 3) for an intermediate one.

Berger (1973) also proved that Sellers' model likewise implies that during the present epoch, three climatic regimes may prevail. The first corresponds to actual observed climatic conditions; the second to a colder climate with temperature 10–15°C below the present one, and the third, to very low negative temperatures at all latitudes (white earth).

Schneider and Gal-Chen (1973) studied climatic stability on the basis of semiempirical models of the thermal regime, supplemented by a consideration of the instability of these processes. For this purpose another term was added to the heat balance equation of the earth-atmosphere system that describes the instability of the thermal regime:

$$m \frac{\partial T}{\partial t} = \sum M_i, \qquad (2.30)$$

where $\partial T/\partial t$ is the variation of the average temperature at the earth's surface with time, m the coefficient of the thermal inertia of the earth-atmosphere system, and $\sum M_i$ the algebraic sum of the heat balance components for the earth-atmosphere system.

This work has confirmed the conclusion on the high sensitivity of the atmospheric thermal regime to small variations of the solar constant, and at the same time proved that when the inflow of solar energy is steady, the thermal regime returns to its initial condition even after very big drops in the average global temperature, probably reaching $-18°C$. Total glaciation of the Earth is possible only when the average temperature drops even more. Thus, the thermal regime of the earth-atmosphere system is highly stable relative to the fluctuations of the mean global temperature that are not caused by long-term variations of the heat inflow.

An analysis of semiempirical models of the atmosphere's thermal regime was carried out by Held and Suarez (1974). They used empirical parameters determined from satellite observations of the outgoing longwave radiation, their results being similar to ours. Held and Suarez have confirmed the finding that the feedback between the thermal regime and the surface area of polar ice strongly enhances the sensitivity of the ice-cover boundary to variations of the solar constant and other climatic factors. They have also confirmed the existence of "critical latitudes" beyond which the polar ice expands equatorward under constant external climatic factors.

Held and Suarez analyzed the effect on climate stability of the additional term corresponding to the horizontal heat redistribution caused by macroturbulent processes. Thus it was possible to obtain a continuous distribution of the average latitudinal air temperature for concentric distribution of the solar ice relative to the poles.

A semiempirical model of the atmospheric thermal regime was developed by Gandin, Il'in and Rukhovets (1973). They constructed a model of air temperature distribution along latitude and longitude and averaged over the altitude.

2 THE GENESIS OF THE CLIMATE

By using the heat balance equation of the earth-atmosphere system these authors have considered the horizontal heat transfer, assuming that it was determined by microturbulent exchange. It was also assumed that the coefficient of turbulent heat exchange did not depend on the coordinates and that the outgoing long-wave radiation was determined as a function of temperature according to the above-given empirical formula. The effect of polar ice on the atmospheric regime was taken into account in a manner similar to that used in our model.

On the basis of empirical albedo values, the authors found the distribution of the mean annual air temperature over the northern hemisphere to be close to observational data.

Later on, the variations of the air temperature and the ice boundary were studied for a changing solar constant. It was found that when it increases by several tenths of a per cent, the polar ice melts completely, and when it decreases by the same amount, a total glaciation of the earth takes place, accompanied by a drastic drop in the average global temperature (white earth).

These authors have pointed out that the high sensitivity of the results to small variations of the solar constant, depended partly on the choice of the albedo values of the earth-atmosphere system with or without ice. By narrowing the gap between these values, it is possible to obtain the critical value of the solar constant corresponding to the total glaciation of the earth, which is close to the results given above.

Since the empirical values of the albedo are determined with a certain error that can not noticeably affect the results of calculations of the critical values of the solar constant, corresponding to the total glaciation of the earth and to the complete melting of the polar ice, it is obvious that all the calculations of these critical values are approximate and will be made more exact as the albedo is determined with more precision and the models of the thermal regime improved.

Nevertheless, the agreement of the results obtained from all semiempirical models that take into consideration the feedback between the temperature field and the polar ice, should be noted. It follows from all the models that a decrease of the solar constant of at most several percent can cause complete glaciation of the earth. As mentioned above, this conclusion testifies to the very high sensitivity of the current climate to changes in climatic factors.

Besides the above works, semiempirical theories of the atmospheric thermal regime were discussed in many studies recently published (Dwayr, Peterson 1973; Gordon and Davis, 1974; Hantel, 1974; Suarez, Held, 1975; North, 1975; 1975a; Cess et al. 1976). Interest of many authors in these models shows that they appear to be useful means for studying the climate genesis.

Let us compare the sensitivity of the atmospheric thermal regime to the change of solar constant obtained from semiempirical models with the results of similar calculations based on more general theories of climate.

Of particular interest is the study by Wetherald and Manabe (1975) of modern climate sensitivity to a change in incoming radiation.

In contrast to the above mentioned works that used the semiempirical models of the average temperature distribution, Wetherald and Manabe used a three-dimensional model of the general theory of climate. This model took into account all the details of the dynamic processes in the atmosphere, as well as the effect of the water phase transitions on the thermal regime and the feedback between the snow sheets, polar ice, and air temperature.

Wetherald and Manabe found that a 4% decrease in the solar constant causes the expansion of the snow and ice covers over the entire earth. A 2% increase of the solar constant leads to a recession of the ice sheet and to a 2°C increase of the mean annual air temperature at the earth's surface at low latitudes, and to as much as a 10°C increase at high latitudes (80°N).

For comparison we give the temperature variations curve calculated according to our semiempirical model for a 2% increase in the solar constant (Fig. 10).

In these calculations (as well as in those on the basis of which Fig. 27 is plotted) the albedo values of the earth-atmosphere system were obtained from satellite data.

It is noteworthy that Wetheral and Manabe's results agree with the main conclusions implied by semiempirical models on the high sensitivity of the current climate to small variations of the heat influx at the

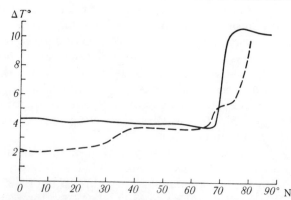

Fig. 10. Effect of a 2% increase in the solar constant on average latitudinal air temperature. Solid line, results obtained from the semiempirical model; broken line, Wetherald-Manabe results.

outer boundary of the atmosphere. Note also the agreement of conclusions on the much higher temperature increase at high latitudes as compared to low latitudes with increasing heat inflow.

The results obtained in these works have considerably changed our conception of the unambiguity and high stability of present-day climate. As we shall see in the following chapters, it is impossible to explain the chief regularities of climatic changes without taking these results into consideration.

3 Contemporary Climatic Changes

3.1 The Warming During the First Half of the Twentieth Century

Secular Temperature Trend. The general climate changes that took place in the first half of the twentieth century were described in Chapter 1. According to investigations carried out by Rubinshtein (1946, 1973), Rubinshtein and Polozova (1966), Mitchell (1963), Lamb (1966) and others, these climatic changes varied in different regions and from season to season.

In particular, it follows from the works of Rubinshtein and Polozova that the mean anomalies of monthly air temperature at a given point vary greatly during successive months, these variations often being of an irregular nature. The relation between temperature anomalies at various points weakens rapidly with distance.

It is possible that the time and space structure of temperature anomalies is caused to a certain extent by random fluctuations of atmospheric circulation. Averaging temperature anomalies over a period of time for certain stations or limited regions does not entirely exclude the effect of such fluctuations, thus making it difficult to reveal the regularities in climatic changes.

For a study of global fluctuations of climatic conditions, the large-scale averaging of the air temperature anomalies is of great importance. This approach is based on a number of physical considerations.

As we know from Chapter 2, there is a linear relation between the long-wave radiation emitted into outer space and the air temperature at the earth's surface. Therefore the averaging of temperature anomalies is equivalent to averaging the outgoing radiation anomalies.

As a rule, the anomalies of the outgoing radiation in limited regions are generally compensated to a great extent by averaging over the surface area, and their effect on the global climate is limited. Climatic conditions are strongly affected by radiation anomalies over large regions and especially by global anomalies.

The algebraic sum of this anomaly and that of the short-wave radiation absorbed by the earth-atmosphere system determines the sign of the air temperature variations over the planet as a whole.

Average air temperature anomalies for large areas were calculated by Mitchell (1961, 1963) on the basis of air temperature observations carried out by meteorological stations in various geographical regions.

For a more precise determination of mean anomalies we have used maps of air temperature anomalies and not the observational data of separate stations. Such maps compiled at the Main Geophysical Observatory give the distribution of the average monthly temperature anomalies for each month beginning with 1881 through 1960 in the northern hemisphere, with the exception of the equatorial zone, for which there is not enough information referring to the first part of this period.

In Fig. 11, the secular variations of the air temperature anomalies are presented as calculated by Spirina, for the extraequatorial zone of the northern hemisphere and for the 70–85°N latitude zone. All the data were averaged over running five-year periods.

From this figure it follows that at the end of the nineteenth century

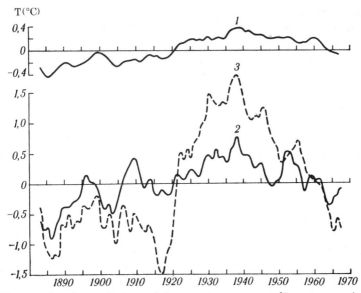

Fig. 11. Secular variations of air temperature anomalies over running five-year periods. (1) Anomalies of the average annual temperature of the northern hemisphere; (2) temperature anomalies of the 70–85°N latitude zone during the warm half-year; (3) the same during the cold half-year.

a warming began in the northern hemisphere which reached its first and weak maximum during the last years before the turn of the century. It was followed by a slight temperature drop which soon took an upward leap. This leap greatly accelerated during the cold period at the end of the 1910s and the beginning of the 1920s. The positive temperature anomaly reached its maximum at the end of the 1930s, while in the 1940s a cooling process began that accelerated in the 1960s. In the 1960s the mean temperature in the northern hemisphere reached the level of the 1910s. Thus the cooling of the last decades has compensated for approximately half of the temperature increase that had been taking place since the end of the nineteenth century.

It appears that the secular temperature trend in the extraequatorial zone of the northern hemisphere follows that of the air temperature at the earth's surface for the entire planet. The data available for the southern hemisphere are more limited than for the extratropical latitudes of the northern hemisphere and show that in the equatorial zone and the extratropical latitudes of the southern hemisphere, the variations of the average air temperature were weaker than in the extratropical zone of the northern hemisphere. Nevertheless, the nature of these variations in most of the zones for which data exists apparently coincided with the variations in the zones covered by numerous observations (Mitchell, 1963).

Figure 11 shows that secular temperature variations increase with latitude, the air temperature during the cold period changing more drastically than during the warm period, especially at higher latitudes. The general picture of the thermal regime variations presented by Fig. 11 is similar to that obtained by Mitchell.

It is noteworthy that in addition to regular temperature variations over long periods, Fig. 11 reveals numerous short-term temperature fluctuations, which reflect the effect of unstable circulation processes that have not been entirely excluded by averaging over space and time.

Due to the fact that temperature anomalies of various seasons differ comparatively little at low latitudes, the average meridional temperature gradient changed less during the warm season than during the cold season of the year.

Secular variations of anomalies of the mean meridional temperature gradient are presented in Fig. 12 (Budyko and Vinnikov, 1973). The anomalies were determined by the root-mean-square method on the basis of the average latitudinal temperatures for every 5°-zone in the interval from 25 to 70°N and subjected to five-year running averaging.

It follows from the figure that from the 1880s to the 1930s the meridional temperature gradient kept dropping. During this period there were two comparatively short intervals when the meridional

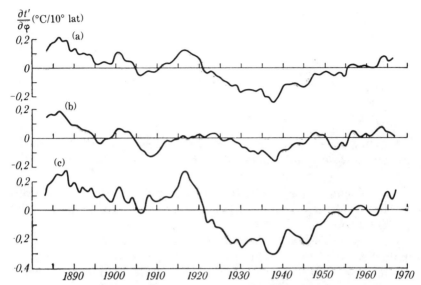

Fig. 12. Secular variations of meridional air temperature anomalies in latitude zone, 70–85°N. (a) mean annual anomalies; (b) mean anomalies for the warm half-year; (c) mean anomalies for the cold half-year.

gradient rose—during the first years of the twentieth century and during the second half of the 1910s.

Beginning with the latter half of the 1930s, the meridional gradient began to climb again, its anomalies by the end of the 1960s reaching the same level as at the turn of the century.

Secular Variations of Precipitation. Along with temperature variations during the warming period, notable changes in the amount of precipitation took place in several regions.

The secular variations of precipitation during the cold part of the year (from November to March), in the steppe and forest-steppe zone of the U.S.S.R. shown in Fig. 13, are based on data compiled by Grigor'eva. The precipitation is calculated for running five-year periods and is based on observational material obtained at 21 stations. From a comparison of Figs. 11–13 we can see that when the average temperature rises and the meridional temperature gradients drop, there is a tendency toward a decrease in precipitation in the regions of unstable moisture.

Drozdov and Grigor'eva (1963) have noted that during the epoch of maximum warming (1930s), widespread droughts in the regions of insufficient moisture over the territory of the U.S.S.R. and North America were more frequent than in the previous and following decades. As

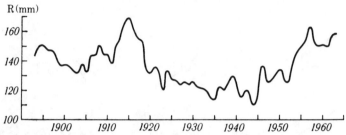

Fig. 13. Secular variations of total precipitation during cold half-year in steppe and forest-steppe zones of the U.S.S.R.

we know, during these years, the level of the Caspian Sea dropped drastically by about 170 cm owing to the decrease of precipitation in the Volga River basin.

Shnitnikov (1969) and others have noted that the fluctuations of the Caspian Sea level are similar in many respects to those of several European lakes, and therefore reflect the large-scale anomalies in the regime of atmospheric precipitation.

It may be assumed that to a large extent the secular variations of precipitation are caused by the variation of the meridional temperature gradient which affects the nature of the atmospheric circulation.

As the meridional temperature gradient decreases, so does the intensity of the water-vapor transfer from the ocean to inland regions, and conversely.

This conclusion is well confirmed by observational data. Figure 14 shows the dependence of the precipitation during the cold period of the year in the steppe and forest-steppe zones of the U.S.S.R. on the meridional temperature gradient for running five-year periods from 1891 to 1960. It follows from this figure that the correlation coefficient is 0.78. The precipitation over this territory varied by 50% from its mean value during various five-year periods. It is obvious that such variations in the amount of precipitation over a vast territory have a great influence on streamflow parameters and that agricultural crop yields depend considerably on precipitation in the regions of insufficient moistening.

The fluctuations of the Caspian Sea level are a good illustration of the dependence between the amount of precipitation and the river discharge. The drastic drop of its level in the 1930s was the consequence of a precipitation decrease over most of the European part of the U.S.S.R. During the same period the meteorological conditions for agriculture over the greater portion of Eurasia and North America also worsened.

It is interesting to compare the changes in humidity conditions in

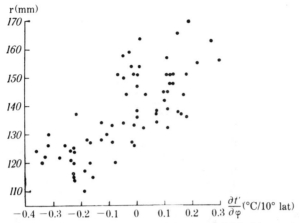

Fig. 14. Dependence of precipitation during cold half-year in steppe and forest-steppe zones of the U.S.S.R. on meridional temperature gradient in 25-75°N latitude zone.

the middle latitudes of the continents in the eastern and western hemispheres. In order to study these variations an analysis was made of the wheat crop deviations from its mean trend in the U.S.A. for separate five-year periods (1910-1971). It follows from these data that during individual intervals the crop changed by as much as 16% as compared to the norm, and that this was caused mainly by the fluctuation of humidity.

Since such data are not available for vast territories of Eurasia, we had to rely on the information concerning the Caspian Sea level in order to study humidity conditions.

In Fig. 15 the wheat crop anomalies in the U.S.A. are compared to the fluctuations ΔH of the Caspian Sea level during the same period. These two parameters seem to correlate comparatively well with a coefficient of 0.74. Here we find confirmation that this mechanism of precipitation change exercises a similar influence on the humidity regime of middle latitudes on different continents as well.

This conclusion is likewise confirmed by direct measurements of precipitation in North America. Landsberg (1960) compared the air temperature and precipitation in the U.S.A. for two 25-year periods: 1906-1930 and 1931-1955. During the second period the average annual air temperature at the earth's surface was 0.5° higher than during the first.

For a comparison of precipitation amounts during these periods, the choice of the years was not a fortuitous one, as a sharp change in the meridional gradient took place not at the boundary of the given periods,

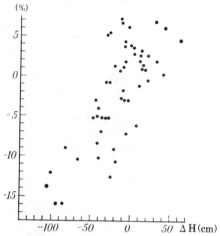

Fig. 15. Correlation between humidity conditions in Eastern Europe (level fluctuations of the Caspian Sea) and in North America (relative wheat crop anomalies in the U.S.A.).

but around 1920. Nevertheless, Fig. 12 shows clearly that the meridional temperature gradient of the second period was lower than that of the first.

As noted by Landsberg, the annual total precipitation over the greater part of United States territory was lower in the second period, which agrees with the above conclusion.

A detailed investigation of the dependence of precipitation on the variations of the meridional temperature gradient was carried out by Drozdov and Grigor'eva (1963, 1971), who found that though at high latitudes the general pattern of precipitation during the warming and cooling periods is comparatively complex, in regions of insufficient humidity in middle latitudes, a tendency nevertheless prevails toward an increase of precipitation as the temperature drops in the Arctic. A similar conclusion could be drawn from the study by Lamb (1974). Drozdov and Grigor'eva explained this effect by the intensified transfer of water vapor to intracontinental regions when the temperature difference between low and high latitudes increases.

This mechanism requires special study. It seems that with the growth of the meridional temperature gradient, the subtropical high pressure belt shifts to a lower latitude. This augments the amount of water vapor being transferred from the ocean to the intracontinental regions in middle latitudes.

Methods of studying atmospheric circulation variations based on the standardization of synoptic processes have led to diverse and often contradictory results. This is apparently due to the differences in the indices of circulation intensity applied in various works and also to the fact that the regions under study did not coincide.

It follows from general considerations that the intensity of the westward transfer and other forms of circulation can be expected to increase when the temperature difference between high and low latitudes increases, because this difference is the main source of energy for large-scale atmospheric motions.

In order to study the circulation variations during the warming period of the twentieth century, it is necessary to apply the methods either of the general theory of climate, or of the empirical analysis of the circulation regime on the planet as a whole (or at least over one of the hemispheres).

Polar Ice. As shown by observations, the distribution of polar sea ice has a great infleunce on the atmospheric thermal regime. At high latitudes the air temperature during the cold time of the year over the ice-free ocean surface drops, as a rule, to only several degrees below zero, because the ocean releases much heat into the air. But in the presence of an ice cover, which significantly reduces the heat flow from the ocean to the atmosphere, the temperature of the lower air layer can drop to several tens of degrees below zero.

Thus the migration of the polar sea ice boundary greatly affects the air temperature in related regions, especially during the cold time of the year. Unfortunately, observational data of the sea ice boundary are greatly limited for the current period of climatic change.

Data on the ice boundary for the last century mainly refer to the Atlantic sector of the Arctic and the adjacent seas; similar information on other regions of the Arctic Ocean and Antarctica is available only for the last 20–30 years.

Many attempts have been made to reveal the relation between climatic fluctuations and ice cover in the North Atlantic and the Barents and Kara Seas.

Rubinshtein and Polozova (1966) have found that the amount of ice in the Atlantic sector of the Arctic began to decrease in the 1920s. This process did not stop with the end of the warming period, but continued in the Barents Sea up to the middle of the 1950s, when the amount of ice began to increase gradually. Off the coast of Iceland, the ice had been decreasing since the end of the last century and through the 1940s, then it began to expand. It is obvious that the polar ice which is closely related to the thermal conditions of the upper layer of sea waters at high latitudes, possesses a certain inertia that causes the ice-cover changes to lag behind the fluctuations of the global climate.

An interesting result was obtained by Flohn (Inadvertent Climate

Modification, 1971), in describing the variations of the ice boundary in the Arctic. He calculated the difference in mean latitudinal temperature averaged over the year in the northern hemisphere, between the periods 1931–1960 and 1961–1970. Flohn's diagram implies that this difference is close to zero from the equator to 50°N; it increases sharply in the polar latitudes, reaching its maximum value of 1° between 70 and 80°N, and then drops again. Such a distribution may be related to the shift of the mean boundary of polar ice toward lower latitudes in the 1960s.

3.2 Causes of Climate Warming

Radiation Factors of Temperature Variations. At the beginning of the twentieth century it was already established that the average amount of direct solar radiation that reaches the earth's surface through a cloudless sky can vary noticeably from year to year. These fluctuations are clearly discernible on the curves of secular variations of direct radiation based on actinometric observations. The curves also show considerable changes over longer periods of time, such as several decades (Budyko and Pivovarova, 1967).

It is interesting to compare the secular temperature variations in the northern hemisphere with those of the radiation reaching the earth's surface. For this purpose actinometric observations carried out at several European and American stations between 1880 and 1965, were reduced and an average curve obtained of the secular variations of direct radiation under cloudless conditions (Pivovarova, 1968). In Fig. 16 curve b presents the solar radiation smoothed over running 10-year periods. As we see, the solar radiation had two maxima; a brief one, at the end of the nineteenth century, and a longer one, in the 1930s. Let us compare curve b with curve a representing the secular temperature variations, also smoothed over decades. We find that there is a resemblance between them. Thus, both curves possess two maxima, one at the end of the nineteenth century, and the second and principal one, in the 1930s. At the same time there are certain differences also. For instance, the first maximum is more pronounced on the radiation curve than on the temperature curve. The similarity of curves a and b leads us to believe that radiation fluctuations caused by irregularities in the atmospheric transparency are an important factor of climatic change. In order to clarify this problem it is necessary to calculate temperature variations as a function of the atmospheric transparency for short-wave radiation.

In his studies mentioned in Chapter 1, Humphreys established that the main factor affecting the atmosphere's transparency on a global

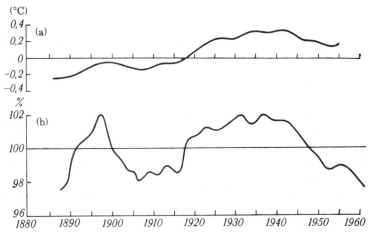

Fig. 16. Secular variations of (a) temperature and (b) direct radiation anomalies.

scale is the presence of comparatively small aerosol particles that remain in the lower level of the stratosphere for a long time.

Humphreys and Wexler assumed that the smallest particles can remain in the atmosphere for several years. These have a very small effect on the long-wave radiation but significantly augment the scattering of short-wave radiation, as a result of which the planetary albedo increases, while the amount of radiation absorbed by the earth as a whole decreases. It should be noted that due to the preferred scattering of the radiation in the direction of the incident beam (Mie effect), the direct radiation decreases much more because of the scattering than does the total solar radiation. Since the earth's thermal regime is affected by the fluctuations of the total radiation, it is necessary to determine the relation between this parameter and the presence of aerosol particles in the stratosphere. For this purpose we can use the methods applied by Shifrin and his colleagues in their studies on atmospheric optics (Shifrin and Minin, 1957; Shifrin and Pyatovskaya, 1959).

This study resulted in a determination of the ratio of the total radiation decrease to the direct radiation decrease for average conditions at various latitudes, when a layer of aerosol particles is present in the stratosphere. The results are given in Table 9.

Shifrin's studies have shown that the values given in this table depend comparatively little on the size of the particles if their prevailing diameters are between $(2-3) \cdot 10^{-2}$ and $(2-3) \cdot 10^{-1}$ μm.

When determining the effects of a change in the amount of direct radiation on the average temperature at the earth's surface, it is neces-

Table 9
Effect of Statospheric Aerosol on the Ratio
of Total to Direct Radiation Decrease

Latitude	90	80	70	60	50	40	30	20	10	0
Ratio	0.24	0.23	0.22	0.21	0.19	0.18	0.16	0.14	0.13	0.13

sary to consider the dependence of mean temperature on the incoming solar radiation. These calculations have been carried out time and again. During recent years it has been found that a 1% change in radiation causes a 1.2–1.5°C change in the temperature for a constant albedo of the earth-atmosphere system (see Section 2.2).

Let us compare the radiation and thermal regimes of the earth during two 30-year periods: 1888–1917 and 1918–1947. From the data shown in Fig. 16, it follows that during the second period, the direct radiation was 2% higher than during the first one. Since, according to Table 9, the mean weighted ratio of the total radiation changes to the direct radiation changes is 0.16, we find that the total radiation during the second period was higher in the northern hemisphere by 0.3%. This corresponds to the increase in the average temperature by approximately 0.4°C. The actual temperature difference for these periods, as determined from the data shown in Fig. 16, is equal to 0.33°C, which agrees well with the results of calculations (Budyko, 1969).

For a more detailed study of radiational changes and their effect on the air temperature, we shall fall back on the model of the thermal regime for various seasons dealt with in Chapter 2.

Figure 17 shows secular variations of the smoothed anomalies of direct radiation for a cloudless sky in the years 1910–1950. This diagram is based on the data obtained at several actinometric stations in Europe and North America scattered over the latitude zone of 40–60°N. In accordance with the above stated hypothesis, let us assume that the variations of direct radiation are caused mainly by the changing transparency of the lower stratospheric layer, due to fluctuations in the concentrations of aerosol particles.

By using the data of Fig. 17, considering the relative optical depth of the aerosol layer, and assuming that the aerosol is homogeneously distributed in the stratosphere over the entire northern hemisphere, we can determine the secular variations of the direct radiation at various latitudes both for the average annual conditions and for separate seasons.

As previously noted, the contribution of aerosol concentration to the changes in total radiation represents a small portion of the variations of direct radiation. By taking into account the dependence of the ratio of total and direct radiations on the altitude of the sun above the horizon, it is possible to calculate the time variations of the total radiation in different latitude zones.

Figure 18 represents the ratio of the variation of total radiation over the northern hemisphere to that of direct radiation at 50°N for the warm and cold half-years, as a function of latitude.

In order to determine the temperature variations corresponding to the above given data on radiation, it is necessary to take into account the relation between the thermal regime and the ice cover.

It is obvious that we cannot rely on the assumption that the ocean-polar ice-atmosphere system is stationary. Due to the large thermal inertia of the ocean and the continental ice cover, this system can be considered stationary only over long periods of time. Therefore, the assumption of its being stationary is not applicable to the study of

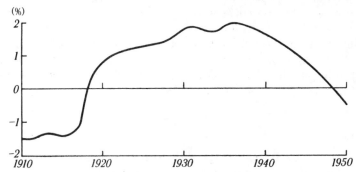

Fig. 17. Secular variations of the direct radiation anomalies during 1910–1950.

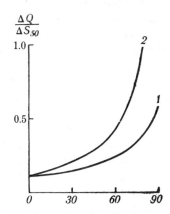

Fig. 18. Ratio of variation of total radiation over northern hemisphere to that of direct radiation at 50°N. (1) Warm half-year; (2) cold half-year.

current climatic fluctuations that have been going on for only a few decades.

At the same time, we cannot ignore the changes of polar ice regime which are connected with climatic variations. For instance, observational data show that the warming of the Arctic has led to an approximately 10% shrinkage of the surface area of sea ice (Ahlman, 1953; Chizhov and Tareeva, 1969).

It is very difficult to carry out precise calculations of nonstationary processes in the ocean-polar ice-atmosphere system, especially since the heat exchange mechanism between the surface and deeper layers of the ocean is not very well known. Therefore, it seems helpful to assume that for a given time period, there is a certain relation between the surface area of polar ice and the climatic factors involved. By determining these parameters from empirical data, we can use these relations as a supplementary equation in our model of the thermal regime, instead of the expression relating the surface area of the ice to the elements of the thermal regime under stationary conditions.

It follows from the semiempirical theory of the thermal regime that the relative change in the ice surface is approximately proportional to the radiation change. This relation is valid for comparatively small surface variations and can be presented in the following form:

$$-\frac{\Delta p}{p} = \mu \frac{\Delta Q_p}{Q_p}, \qquad (3.01)$$

where $\Delta p/p$ is the relative change of the polar ice surface area; $\Delta Q_p/Q_p$ is the relative variation of the total planetary radiation, and μ a dimensionless coefficient.

For current climatic changes the coefficient μ can be determined from empirical data on the variation of the total radiation and the ice cover during the Arctic warming period. Figures 16 and 17 imply that during this epoch the direct radiation at 50°N exceeded that of the previous period by about 2%.

Assuming that the surface of polar ice shrank during the same epoch by 10%, we find that $\mu = 40$. This is much less than the similar coefficient which can be derived on the basis of the semiempirical theory of the thermal regime for stationary conditions.

It must be pointed out that this coefficient depends on the time period to which it refers and that its precision is not too high. Numerical experiments have shown, however, that the error in the determination of coefficient μ does not strongly affect the calculated air temperature distribution, which justifies the idea of using an approximate value of this coefficient.

By using Fig. 17 and the described numerical model of the thermal regime, where the relation (3.1) replaces the dependence of the ice boundary on the air temperature, we can calculate temperature variations at different latitudes during a given period. These results are presented in Fig. 19 by broken curves which appear to be quite close to the observed temperature variations smoothed over 10-year periods (continuous curves).

It may be noticed that the calculated values somewhat precede the observed values. The lag of observed curves is apparently due to the inertia of the ocean-polar ice-atmosphere system, though in this case it is not too great. In the thermal regime model applied here the derived empirical parameters do not take into consideration temperature variations, and therefore its results do not depend on the applied experimental data.

From an analysis of temperature variations at different latitudes, it follows that in the northern hemisphere the main reason for temperature increase in the 1920s and 1930s was the increase of total radiation reaching the earth's surface. The temperature increases during the warm half-years that were revealed in the calculations and determined from observations, differed very little at different latitudes.

In the cold half-year temperature fluctuations at low and middle

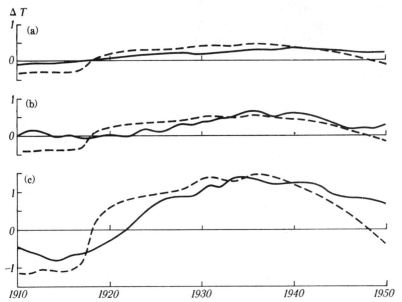

Fig. 19. Secular variations of air temperature anomalies. Solid curves, observational data; broken curves, calculated data.

latitudes differ very little from those of the warm half-year. But at high latitudes, mainly in the 70–80° zone, temperature fluctuations are greatly augmented.

In this case temperature variations depended very little on the fluctuations of solar radiation in the same season, because during the cold half-year the radiation at high latitudes is very low and scarcely affects the atmospheric thermal regime. The primary cause of temperature variations in this case was the changing area of the polar sea ice that resulted in a considerable rise of air temperature during the cold half-year.

This effect on the temperature in the warm half-year is comparatively small; it also diminishes rapidly in the cold period as the distance from the 70–80°N latitude zone grows.

This conclusion agrees well with the earlier conception of the influence of polar ice on the air temperature, according to which the presence of ice causes a sharp decline in the air temperature at high latitudes during the winter, exercises a smaller effect in the same zone during the summer, and a still smaller influence on the air temperature in middle and low latitudes (Budyko, 1971).

Role of Volcanic Eruptions. Savinov (1913), Kimball (1918), and Kalitin (1920) as well as other authors have established that explosive volcanic eruptions result in a drastic reduction of solar radiation reaching the earth's surface.

In such cases the average amount of direct radiation for vast territories may drop 10–20% in the course of several months or years. An example is given in Fig. 20 which illustrates the average monthly values of direct radiation under a clear sky, relative to normal value

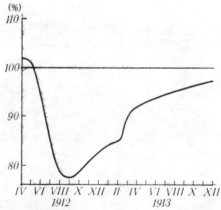

Fig. 20. Average monthly values of direct radiation after volcanic eruption relative to normal values.

after the eruption of the Katmai volcano in Alaska. This curve, based on observations carried out at several actinometric stations in Europe and America, shows that the atmospheric aerosol caused a more than 20% drop in the direct radiation during specific months.

In certains areas the radiation drop was even greater. For instance, in Pavlovsk, a suburb of St. Petersburg (Russia), situated at an enormous distance from Alaska, the solar radiation for half a year was 35% below the norm. Similar effects were produced by the eruption of the Krakatao volcano in Indonesia in 1883. In both cases, after the eruptions, anomalous optical effects in the atmosphere were observed over great territory, thus confirming the global character of changes in radiation regime resulting from the spread of the aerosol in the stratosphere.

The establishment, mainly in the 1950s, in the course of preparations for the International Geophysical Year, of a world-wide network of actinometric stations greatly facilitated studies of the effects of volcanic eruptions on the solar radiation regime. Since then, the first major explosive eruption occurred in March, 1963 (the Agung volcano on the island of Bali, in Indonesia). The effects of this eruption on the radiation regime were studied with far greater thoroughness than was ever possible in the past.

Shortly after this eruption, the effect on incoming radiation was detected in various regions of the planet (Burdecki, 1964; Flowers and Viebrock, 1965; Dyer and Hicks, 1965, 1968; Budyko and Pivovarova, 1967).

Budyko and Pivovarova reduced observations of direct noontime radiation, carried out at several actinometric stations of the Soviet Union during 1957 to 1966.

They established that the average monthly values obtained at 22 stations situated between 40° and 68°N, registered very little change from 1957 to November, 1963. However, in December, 1963, these values sharply declined (Table 10). A systematic change in these values since

Table 10
Deviation of Direct Radiation from Average over U.S.S.R. (in %)

Year	I	II	III	IV	V	VI	VII	VIII	IX	X	XI	XII	Annual
1963	−1	3	−2	2	2	2	2	1	0	−1	−2	−10	0
1964	−10	−3	−6	−7	−2	−1	−3	−2	−3	−6	−8	−16	−5
1965	−15	−7	−11	−5	−4	−3	−3	−2	−5	−2	−9	−12	−6
1966	−10	−9	−7	−4	−2	−1	1	0	−2	−4	−7	−7	−4

the end of 1963 can be seen along with certain variations of direct radiation from month to month caused apparently by the instability in the atmospheric circulation.

There was a big drop in the intensity of direct radiation in the winter as compared with the summer months, which means that it was caused by changes in the atmospheric transparency and not by the fluctuations of the solar constant.

Similar calculations were carried out for other parts of the globe. The difference between the intensity of direct radiation at noon in 1964 and 1958 was determined on the basis of observations conducted at several stations outside of the Soviet Union while the transparency of the atmosphere in 1958 was assumed to be close to the average value for the end of the 1950s and the beginning of the 1960s.

These results are given in Table 11. They clearly show the same attentuation of direct radiation in 1954 over Eastern Europe, North America, and the Central Pacific.

The absence in this case of a noticeable annual variation in the intensity difference can be explained by the fact that all the stations listed in Table 11 are located at comparatively low latitudes where the average midday altitudes of the sun are not very different all year round.

Observations of scattered radiation show that after 1963, the decrease in direct radiation over the U.S.S.R. was accompanied by a sharp increase of scattered radiation. This conclusion follows, for example, from Table 12 which presents the relative variations of the scattered radiation observed at noon under clear sky conditions, at 22 Soviet stations.

We find by comparing the absolute values of the decreased direct radiation and the increased scattered radiation over the territory of the

Table 11
Deviation of Direct Radiation in 1964 from that of 1958 (in %)

Station	I	II	III	IV	V	VI	VII	VIII	IX	X	XI	XII	Annual
Porto, Lisbon, Faro (Portugal)	−7	−10	−6	−11	−9	−9	−10	−7	−12	−10	−12	−14	−10
Blue Hill, Albuquerque (N. America)	−12	−5	−5	−3	−4	−4	−4	−4	−6	−5	−9	−4	−6
Mauna Loa (Hawaii)	−	−4	−	−7	−6	−	−6	−4	−6	−6	−6	−4	−6

Table 12
Deviation of Scattered Radiation From Average over U.S.S.R. (in %)

Year	I	II	III	IV	V	VI	VII	VIII	IX	X	XI	XII	Annual
1964	10	7	12	22	17	6	19	6	3	9	11	25	14
1965	20	14	24	17	22	25	12	6	8	9	11	12	14

Soviet Union, that the total radiation changed very little during these years. Although it is difficult to determine the amount of this variation from empirical data, it can be calculated by taking into account the variations of direct radiation.

Dyer and Hicks (1968) studied the effect of the Agung eruption on the direct radiation observed at several dozen actinometric stations situated in the northern and southern hemispheres. They found that the aerosol particles spread over the entire globe, from the south pole to high latitudes of the northern hemisphere. This did not take more than several months and their concentration remained high for about two years; for the Soviet Union this period was longer. It is possible that only comparatively small particles reached the middle latitudes of the northern hemisphere and that it took them longer to settle. Dyer and Hicks describe the properties of the aerosol cloud which formed after the Agung eruption. The average altitude of this cloud was 15–20 km, the mean radius of the aerosol particles varied from 0.5–1.0 μm during the first month of its existence, to 0.1–0.15 μm one year later. Dyer and Hicks assumed that the cloud began to form at an altitude of 22–23 km and that its particles later settled at a speed corresponding to their sizes.

Considering the great influence of volcanic eruptions on the radiation regime, we can assume that the radiational increase at the end of the nineteenth century was the result of atmospheric clearing of the aerosol caused by the Krakatau eruption. The succeeding drop in radiation was the consequence of the eruption of the Mont Pelée and other volcanoes (anomalies began to decrease on the curves of secular variations of radiation in Figs. 16 and 17 before the time of these eruptions, but this is due to the 10-year smoothing effects). The increase in radiation during 1915–1920 was apparently due to the transparency increase caused by the settling of aerosol after Katmai eruption, after which there were no intensive eruptions for a long time.

It is interesting to analyze the radiational drop that began in the 1940s. A number of scientists have expressed the opinion that this was the result of industrial pollution (Davitaya, 1965) as well as nuclear tests. It could also be due to eruptions from the Spur volcano (Alaska) in

1953, the Bezimyannaya Sopka (Kamchatka) in 1956, and others. This problem is dealt with in greater detail in Chapter 6.

Several empirical works show that after major eruptions the average temperature at the earth's surface was lowered by several tenths of a degree during periods varying from several months to several years (Humphreys, 1929; Mitchell, 1961, 1963).

Let us consider the effect of these radiation changes on the temperature. We can make a theoretical estimate by taking into account the heat conductivity and the specific heat of ocean water. However, since we are interested in only an approximate estimate of this effect, we shall use a simple relation between radiation and temperature variations that can be established empirically from the annual variations of the parameters.

We shall assume that the rate of the temperature variations at the earth's surface is proportional to the difference between the present temperature (T) and the long-term average temperature (T_r), i.e.,

$$\frac{dT}{dt} = -\lambda (T - T_r), \tag{3.2}$$

where λ is the proportionality factor.

If the temperature at some initial time is designated by T_1, then from (3.2) we obtain

$$T - T_r = (T_1 - T_r) e^{-\lambda t}. \tag{3.3}$$

For an approximate estimate of λ let us use the data on the annual variations of solar radiation and temperature in the northern hemisphere neglecting in these calculations the interaction between the climatic factors in the northern and southern hemispheres.

Assuming that the ratio of the solar radiation at the outer limits of the atmosphere in the northern hemisphere during the warm half-year (April–September), to the mean annual radiation is equal to 1.29, we find from the above formula, relating the temperature with the radiation changes, that this change in radiation could lead to a temperature increase of approximately 40° in the case of zero heat inertia. The observed difference between the average temperature in the northern hemisphere for the warm half-year and that for the entire year is equal to 3.5°C. Assuming here that $T - T_1 = 3.5°C$ and $T_r - T_1 = 40°C$ for $t = \frac{1}{4}$ year, we find from (3.3) that $\lambda = 0.4 \ yr^{-1}$.

With this in mind, let us find the change in average temperature one year after a volcanic eruption that led to a 10% drop in direction radiation. In this case the drop in the total radiation would be 1.5% and the decrease in T_r about 2°C. From (3.3) we find that the temperature variation after the eruption is equal to several tenths of a degree. This result agrees well with the annual average temperature anomalies registered after large eruptions of an explosive nature.

A more detailed calculation of temperature changes after volcanic eruptions was carried out by Borzenkova (1974) who also applied the semiempirical model of the atmospheric thermal regime and the given formula for finding the average temperature variations at the earth's surface. Borzenkova found that after a volcanic eruption the temperature drop in the northern hemisphere increases with latitude both in the warm and cold half-years.

Oliver (1975) and Sagan with co-workers (Pollack et al., 1975) studied the volcanic eruption effect on climate. The results of their calculation are in good agreement with the empirical data on the mean temperature variations. These studies corroborate the assumption that contemporary climatic changes depend considerably on the level of volcanic activity.

Detailed empirical studies of the effect of volcanic eruptions on the air temperature, carried out by Pokrovskaya (1971) and Spirina (1971), have revealed regularities in the effect of a single eruption on the thermal regime. It has been established by these works that for several years after powerful eruptions, there is a reduction in air temperature during the warm part of the year, this reduction being the greatest in the northern part of the middle latitudes. During the cold seasons temperature variations after eruptions are more complicated: the temperature usually drops in the polar zone and often increases at middle latitudes. As a result, the average annual temperature takes a bigger drop in high latitudes as compared with middle latitudes.

The abovementioned regularities, according to Pokrovskaya, are the result of the predominant effect of radiation factors on climatic conditions during the warm seasons, when the increase in the optical depths of the aerosol layer enhances the temperature drop in the polar regions. During the cold season, the main factors of temperature distribution are the fluctuations in atmospheric circulation caused by variations of the heat inflow to various latitudes.

Tables 10 and 11 show that at middle latitudes the relative effect of aerosol particles on radiation is much smaller during the warm than during the cold season, the reason for this probably being the noon solar altitude.

Since the radiation in middle latitudes during the cold season is much lower than during the warm season, the annual variations of the absolute radiation are significantly less than the relative values given in Tables 10 and 11; nevertheless, a comparatively large decrease in radiation during the cold period cannot but affect other components of the heat balance of the earth-atmosphere system. It is possible that it influences mainly the heat exchange in the upper layers of the ocean, whose great thermal inertia causes variations of the summer temperatures discovered by Pokrovskaya and Spirina.

These scientists have shown that the variations in the atmospheric thermal regime caused by a single volcanic eruption are not stationary and that the temperature drop in this case is only a part of the changes that would have taken place under stationary conditions.

The effect of volcanic activity on temperature changes occurring over longer intervals lasting several decades was studied by Lamb (1970), who compared temperature anomalies for various time intervals with the index representing the average drop in atmospheric transparency after eruptions.

Lamb has expressed this index in relative units which describe the effect of volcanic eruptions on the atmospheric transparency in terms of the effect of the Krakatau eruption of 1884.

A highly negative correlation coefficient of -0.94 was obtained between the temperature anomalies in the northern hemisphere (from the equator to 60°N) for the period of 1870–1959, averaged over 10-year intervals, and the described index, which proves the close correlation between the volcanic activity and temperature fluctuations. Lamb pointed out that by the end of this period the relation was disturbed. The reasons for this will be analyzed in Chapter 6.

For certain regions this correlation appears to be less close, and this seems only natural considering the great influence of atmospheric circulation changeability on temperature anomalies in limited regions. From the point of view of our theory on the mechanism of modern climatic changes, the relation discovered by Lamb between the amount of sea ice off the coast of Iceland and his index of atmospheric opacity, is of great interest. The correlation coefficient between these parameters for 10-year intervals, from 1780 to 1959, was found to be 0.61, which is rather high in view of the fact that Lamb was comparing very approximate estimates, especially for the first part of the considered period.

Aerosol Layer in the Stratosphere. In order to study the mechanism by which volcanic eruptions influence the climate, it is necessary to make a more thorough study of the dependence between climatic conditions and atmospheric aerosol. The atmosphere includes not only water droplets and ice particles of clouds and fog, but also a large amount of solid and liquid suspended particles of varying chemical composition. The size of these particles ranges from 10^{-2} to 10^{-7} cm.

The greater portion of the total aerosol mass is comprised of the so-called large particles with radiuses ranging from 10^{-5} to 10^{-4} cm and gigantic particles measuring over 10^{-4} cm.

Available estimates show that from 800 to 2200 million tons of matter enter the atmosphere annually, as a result of natural processes, from which aerosol particles are formed. Another 200–400 million tons of matter is added as a result of human activity. Among the aerosol particles are products of rock and soil weathering, sea salt from the ocean surface, soot and ashes from forest fires and combustion gases, as well as products of chemical reactions of sulphurous gas, hydrogen sulphide, ammonia and other gases entering the atmosphere from the earth's surface (Inadvertent Climate Modification, 1971).

Atmospheric aerosol is concentrated mainly in the lower layers of the trophosphere and its residence time is, as a rule, comparatively short. The largest particles settle down rapidly under the force of gravity, while the smaller particles are carried to the earth's surface by the descending air currents and by precipitation playing an important part in cleansing the air of aerosol.

According to available data, the average residence time of an aerosol particle in the troposphere is about 10 days. Therefore we find that the atmosphere contains approximately 30–70 million tons of aerosol.

In the stratosphere the amount of aerosol is considerably less. It remains there much longer, however, than in the troposphere.

The main mass of stratospheric particles ranges in size from 1 to 0.1 μm (large particles) whose rate of settling under the force of gravity is comparatively small. Since the vertical air motions in the stratosphere are very poorly developed and there is no precipitation, the residence time of aerosol in the stratosphere ranges from several months to several years.

Direct observations of the composition of the stratosphere aerosol initiated by Junge (1963) have proved that this aerosol is made up of droplets of sulphuric acid and also (to a lesser degree) of sulphuric acid salts, basically compounded with ammonia. Most of the stratospheric aerosol is generally concentrated in a layer several kilometers thick, its middle level being at an altitude of 18–20 km. This is known as the sulphate layer, or Junge's layer.

According to Junge, the aerosol particles of the sulphate layer are products of sulphurous gas that enters the stratosphere from lower levels. After photochemical reactions with atomic oxygen in the stratosphere, the sulphurous gas forms sulphur dioxide, from which, as a result of interaction with water vapor, droplets of sulphuric acid are formed.

Recent investigations show that if the source of stratospheric aerosol is outside the tropical latitudes, the aerosol spreads comparatively fast over the pertinent hemisphere but very slowly over the other hemisphere. If the source of aerosol is close to the equator, it spreads over both hemispheres (Karol', 1972).

The particles of stratospheric aerosol settle gradually under the influence of gravitation force and as a result of large-scale air motions which carry them to the troposphere, where they are rapidly washed out by precipitation. The latter mechanism is apparently the main one for large particles, while the former is more effective for gigantic particles, the amount of which in the stratosphere is low due to their rapid descent.

Karol' (1973) asserts that large particles remain in the stratosphere at an altitude of 20–30 km on an average for 20–40 months, while at the tropopause this period is reduced to 6–20 months. These periods vary greatly with the intensity of air exchange between the stratosphere and the troposphere.

In studying the climatic effects of atmospheric aerosol, it is very important to establish the relation between the vertical radiation flows in the atmosphere and the aerosol concentration. This problem has been studied by many researchers beginning with Humphreys.

Humphreys' contention that the atmospheric aerosol generally has comparatively little effect on long-wave radiation has been confirmed by later research.

At the same time it was established that atmospheric aerosol can noticeably affect the flow of short-wave radiation due to the backscattering and absorption of radiation by aerosol particles.

The scattering index for particles within the size range of 10^{-1}–$10°$ μm is strongly prolate in the direction of the incident beam and as a result the direct radiation is more attenuated than the total radiation.

Due to a shortage of experimental data, the absorption of shortwave radiation by aerosol particles is less clear to science than backscattering. The available data show that as a result of absorption by aerosol particles, the depletion of direct radiation is apparently less than that caused by backscattering, although both effects are comparable in magnitude (Gaevskaya, 1972).

A number of estimates could be listed of an influence of atmospheric aerosol mass upon attenuation of short-wave radiation flux. The simplest way to handle this problem is to compare the average climatological value of short-wave radiation flux attenuation due to aerosol with the average mass of atmospheric aerosol.

Let us assume, in agreement with Pivovarova (1968) and other authors, that this decrease in direct radiation due to aerosol comes to about 10%. From the above data it follows that the average mass of atmospheric aerosol is $50 \cdot 10^6$ t, or $10^{-5} g/cm^2$. About half of this amount is made up of gigantic particles; their number density is not high and therefore their influence on the radiation processes is small. Thus a mass of aerosol particles equal to $0.5 \cdot 10^{-6}$ g/cm² reduces the direct radiation by 1%.

If this attenuation was due only to backscattering of radiation by aerosol particles, the corresponding drop of the total radiation under average conditions would have been about 0.15%. If this attenuation was due entirely to absorption by aerosol particles, it would have been equal to 1%. Let us assume that both effects have comparable influence on the direct radiation attenuation. In this case the amount of aerosol causing a 1% drop in the total radiation is equal to about 10^{-6} g/cm^2. This estimate is of course very approximate.

In several works the effects of the aerosol on the attenuation of total radiation was attributed to backscattering. In the first work of this kind Humphreys (1929) found that the radiation dropped by 1% when the aerosol mass M was equal to $0.6 \cdot 10^{-6}$ g/cm^2. Similar calculations were recently carried out at the Main Geophysical Observatory by Novosel'tsev and Vinnikov, who respectively followed the scheme of multiple scattering suggested by Soboliev, and the scheme of a single scattering developed by Shifrin. The values of the mass M obtained by them are $0.4 \cdot 10^{-6}$ and $0.6 \cdot 10^{-6} g/cm^2$, respectively. A somewhat larger value of M = 10^{-6} g/cm^2 was obtained by Barrett (1971).

The value of M can also be found by calculating the backscattering according to Yamamoto and Tanaka's model (Inadvertent Climate Modification, 1971). These data contend that $M = 1.3 \cdot 10^{-6} g/cm^2$.

The relation between the aerosol mass and the short-wave radiation can also be estimated by Ångström's formula (1962) which shows the connection between the albedo of the earth-atmosphere system and the coefficient of optical opacity for an average mass of aerosol; here $M = 0.6 \cdot 10^{-6} g/cm^2$.

Hence these and similar calculations imply that M is on the average between $0.4 \cdot 10^{-6}$ and $1.3 \cdot 10^{-6} g/cm^2$.

Since the value of M is much less than aerosol mass variations we come to the conclusion that fluctuations of atmospheric aerosol in space and in time can noticeably affect the flow of short-wave radiation reaching the surface of the earth. These fluctuations apparently have a strong effect on the thermal atmospheric regime.

It is instructive to study the connection between the fluctuations of the atmospheric aerosol concentration and the volcanic activity. Circumstantial evidence here is provided by the established decrease of direct solar radiation after major volcanic eruptions as a result of the considerable increase of the aerosol concentration.

There is also direct evidence. Measurements carried out by high-altitude aircraft have shown that the aerosol concentration in the atmosphere over the northern hemisphere increased greatly by the end, as compared to the beginning, of the 1960s. This variation was apparently due to the eruption of Mt. Agung in 1963 and subsequent eruptions of other volcanoes (Inadvertent Climate Modification, 1971).

Causes of Climatic Changes. We may now draw a conclusion in the light of the data given in this chapter about the causes of climatic changes that took place in the first half of our century. The warming which reached its peak in the 1930s apparently resulted from an increase in the transparency of the stratosphere which led to an increase in the solar radiation flow to the troposphere (the meteorological solar constant). This, in turn, led to an increase in the average global temperature at the earth's surface.

The air temperature variations of different latitudes and seasons depended on the optical depth of the stratosphere's aerosol and on the shift of the polar sea ice boundary. The subsequent recession of the Arctic ice has led to an additional and noticeable increase in air temperature during the cold season in high latitudes of the northern hemisphere.

These conclusions are confirmed by calculations based on a model of the atmospheric thermal regime, whose results agree well with observational data.

It seems possible that the changes in the stratosphere's transparency that took place during the first half of the twentieth century were related to the volcanic activity, when the products of the eruptions, including sulphurous gas, were ejected into the stratosphere. This conclusion, although supported by abundant observational material, is nevertheless not as obvious, in our opinion, as the previous one.

It should be noted that the given explanation concerns only the main pattern of the climatic fluctuations that took place in the first half of the twentieth century. Parallel with the above-described general regularities, this process exhibited many specific features of climatic variations during shorter time intervals and in certain geographical regions.

It seems to us that these fluctuations were caused to a large extent by changes in the circulation of the atmosphere and hydrosphere that were sometimes of a random nature and other times, the result of autooscillation.

There are grounds for believing that the climatic variations of the last 20–30 years began to depend to some extent on the activity of man. This problem is discussed in Chapter 6.

Although the warming in the first half of the twentieth century had a certain effect on human economic activity and proved to be the largest climatic fluctuation that took place during the epoch of instrumental observations, its scale was negligible as compared to the climatic changes that occurred during the Holocene, to say nothing of the Pleistocene, when vast glaciations evolved.

This warming, however, is of great importance for revealing the mechanism of climatic changes, since it is the only one that has been comprehensively and exhaustively studied by means of reliable instruments at many research centers.

Therefore any quantitative theory of climatic fluctuations must be backed by data referring to the warming of the first half of the twentieth century.

4 Climates of the Past

4.1 The Quaternary Period

Climate of the Quaternary Period. In Chapter 1 it was pointed out that the Quaternary – the last geological period – was marked by a great variability in climatic conditions, especially at middle and high latitudes.

Natural conditions during this period have been studied much more extensively than any earlier ones. Particularly important contributions to the study of these conditions were made by the Soviet scientists Gerasimov and Markov (1939), Markov (1955, 1960), Saks (1953), Rukhin (1962), Velichko (1973), and others. Valuable data on the climate of the Quaternary can be found in the works of Butzer (1964), Emiliani (1955), Erikson (1968), Ewing and Donn (1956), Fairbridge (1967), Flint (1957), Flohn (1963, 1964), Mitchell (1965a), Schwarzbach (1950, 1961, 1968), Zeuner (1959), and others.

Though numerous outstanding achievements have been made in the study of the Pleistocene, many important natural processes of that time are still insufficiently clear; among these is the dating of cooling epochs associated with the spreading of ice covers over land and sea. Thus the question of the total duration of the Pleistocene remains open.

Of signal importance for working out the absolute chronology of the Quaternary period are the methods of isotope analysis, which include the radiocarbon and potassium-argon methods. The first of these gives us more or less reliable results only for the last 40–50 thousand years, i.e. for the final phase of the Quaternary period. The latter method is applicable to periods of longer duration but its precision is much lower than that of the former method.

In view of the major discrepancies in the opinions of various authors on many important climatic features of that time, we shall limit ourselves to a general description of the Pleistocene climate, supplementing the brief information already given on it in Chapter 1.

The Pleistocene was preceded by a long cooling process which was most pronounced in the middle and high latitudes. This process accelerated during the Pliocene, the last epoch of the Tertiary period, when large ice covers originated in the polar zones of the northern and southern hemispheres.

Paleogeographic data reveal that the glaciation of the Antarctic and the Arctic lasted over several million years. At the beginning the ice covers were comparatively small but they spread gradually to lower latitudes and later receded, so that ice ages alternated with interglacial periods.

It is difficult for many reasons to determine the point at which the repeated migration of the ice cover boundaries commenced, but it is generally believed to have been about 700 thousand years ago.

Moreover, researchers have often claimed that this epoch of the active development of widespread glaciation was preceded by a still longer transitional period, the Eopleistocene, which would add more than a million years to the duration of the Pleistocene.

The total number of glaciations must have been very large because the main ice ages, already recognized by scientists in the last century, seemed to have consisted of a series of alternating warm and cool subperiods, the latter of which may be considered ice ages in their own right.

As mentioned in Chapter 1, the total duration of all the ice ages constituted less than half of the Pleistocene epoch, i.e. they were shorter than the comparatively warm interglacial periods.

The scale of glaciation differed greatly in the various ice ages. Some authors are of the opinion that by the end of the Pleistocene the glaciations were more intensive than the earlier ones had been.

The last one—The Würm, or Valdai, glaciation—that took place several tens of thousands of years ago has been studied most intensively. During that epoch, along with the development of continental ice covers, the climate became very dry over vast areas in middle latitudes. This was probably due to the sharp drop in evaporation from the oceans as the sea ice expanded to lower latitudes. As a result, the rate of the hydrological cycle was reduced, leading to a decrease in precipitation over the continents, which in turn limited the further growth of continental glaciation. In addition, the enlargement of the terrain due to the lowering of the ocean level could also have influenced the precipitation regime, as a great quantity of water was trapped in the ice cover. This process also enhanced the continental nature of the climate and led to a reduction in the amount of precipitation over land. There can be no doubt that the permafrost area increased enormously during the last ice age.

Although climatic conditions in the ice-covered zone at that time were in many respects similar to those of today, it is quite possible that the climate in the ice-free zones of the middle latitudes was then quite different from all currently existing types of climate.

This climate was marked by a combination of comparatively low air temperature all year round, relatively high influx of solar radiation, and a small amount of precipitation over a considerable part of this zone. These climates were conducive to the formation of specific landscapes resembling both the tundra and steppe of our times.

During the Würm glaciation, the tropical climate changed little in comparison with the climate of middle latitudes, although there was a certain drop in the average air temperature and a redistribution of the amount of precipitation.

The last glaciation ended 10–15 thousand years ago and this is generally considered the end of the Pleistocene and the beginning of the Holocene, the epoch in which natural conditions began to be affected by the activity of man.

Astronomical Causes of Climatic Variations. It is difficult to point out the causes of past glaciations because presumably no great changes in climatic factors occurred during the Quaternary period. Only two factors can be definitely identified.

The first is the variation in the amount of radiation reaching the earth's surface at various latitudes throughout the year, due to the changing position of the earth relative to the sun. This position depends on the eccentricity of the earth's orbit, the obliquity of the ecliptic plane and the precession of the equinoxes. All these orbital elements change periodically and cause variations in the amount of radiation that can be calculated quite precisely for the past tens of thousands of years, though with mounting uncertainty for more remote times.

The second factor is the variation in atmospheric transparency due to the amount of aerosol produced mainly by volcanic activity. These fluctuations somewhat affect the amount of radiation reaching the earth's surface, which can be estimated on the basis of contemporary actinometric observations. Some idea of the atmospheric transparency in the past is provided by geological data on volcanic activity during various epochs.

It should be noted that the effect of these two factors on the glaciation regime was not clear until recent times. Since these factors exert only a limited effect on the radiation regime of the earth, many authors have expressed doubt as to their possible influence on the development of glaciations.

In order to clarify this problem, it is necessary to apply a numerical model that allows us to calculate the position of the ice sheet boundary as a function of external climatic conditions. Let us first consider the astronomical factors.

For this purpose we have applied the numerical model of the thermal regime for average annual conditions described in Chapter 2 (Budyko, 1968). This model implies that the variations of the radiation regime during the last Würm glaciation could have led to a 1° southward spread of the ice cover in the northern hemisphere, but this is much below the actual value.

In analyzing this result, we have come to the conclusion that the determination of average annual temperatures is not sufficient for estimating the effect of orbital changes on glaciation, because it depends mainly on the thermal conditions during the warm half-year.

In a later study Budyko and Vasishcheva (1971) used the second model described in Chapter 2, which gives the distribution of the average latitudinal temperature for various seasons.

On the basis of this model the approximate position of the average ice-cover boundary was determined by the method of successive approximations for the periods when the incoming radiation in high latitudes during the warm half-year was reduced due to astronomical factors. Data on the radiation regime for such periods were taken from the work of Milankovich (1941).

In this calculation the dependence of the planetary albedo on the polar ice boundary was taken into account by the equation

$$\Delta \alpha_{sp} = \frac{Q_{sif}^N \Delta L_i^N}{Q_{sp} L_0} (\alpha_{si}^N - \alpha_{sif}^N) + \frac{Q_{sif}^S \Delta L_i^S}{Q_{sp} L_0} (\alpha_{si}^S - \alpha_{sif}^S), \qquad (4.1)$$

where $\Delta \alpha_{sp}$ is the variation of the planetary albedo as compared to its present value; ΔL_i is the variation of the zonal area covered by ice in one of the hemispheres, as compared to the present regime; L_0 is the earth's surface; α_{si} is the albedo of the polar ice zone; α_{sif} is the albedo of the ice-free zone which freezes as a result of radiation decrease; Q_{sif} is the radiation at the outer boundary of the atmosphere in the ice-free zone, whose surface is equal to L_i; Q_{sp} is the average planetary value of radiation. The superscripts "N" and "S" indicate the northern and southern hemispheres.

In this calculation the fluctuations in the ocean radiation balance and cloudiness were not considered.

The results of these calculations are given in Table 13. This table shows that the fluctuations caused by astronomical factors can lead to significant climatic variations. Calculations show, however, that although the average planetary temperature changes vary slightly, the shift of the ice-cover boundary is substantial.

According to these calculations which give the current position of the average ice boundary as close to 72°N and 63°S, the maximum shift

Table 13
Climatic Changes During Glaciation Periods

Time (10^3 yr before 1800)	$\Delta\phi_N°$	$\Delta\phi_S°$	$\Delta T°$
22.1 (Würm III)	8	5	−5.2
71.9 (Würm II)	10	3	−5.9
116.1 (Würm I)	11	2	−6.5
187.5 (Riss II)	11	0	−6.4
232.4 (Riss I)	12	−4	−7.1

Note: $\Delta\phi_N°$ and $\Delta\phi_S°$, decrease in mean latitude of the polar ice boundary in the northern and southern hemispheres, respectively, as compared to their present value; $\Delta T°$, variation of average temperature of warm half-year at latitude 65°N.

of the ice boundary during the studied period was equal to 12° in the northern and 5° in the southern hemispheres.

The zone into which the ice cover spreads experiences a considerable drop in temperature. Thus, at 65°N the average temperature of the warm half-year decreases by 5–7°C. These figures apply to temperatures at sea level; it is possible that during the formation of massive continental glaciations, the temperature drop at the ice level can be much greater.

It is interesting to compare the results of Table 13 with paleogeographic data on natural conditions during the ice ages. Such a comparison, however, meets with certain difficulties resulting from the schematic nature of our calculations and our insufficient empirical knowledge of past climatic conditions.

One of the numerous assumptions made in the course of the calculations is that the positions of the ice cover and the thermal regime were stationary during the periods under consideration. It is known, however, that the development of ice covers was actually a gradual process which lagged behind the maximum radiation decrease in high latitudes. Neglecting this effect may cause a certain error in the calculations of the surface occupied by the ice sheet.

A comparison of these results with empirical data is also difficult because of the absence of exact dating of the Quaternary glaciations. The correlation between the data given in the table and the two main epochs of the last two glaciations corresponds to Milankovich's views, which, however, are not shared by all scientists engaged in the study of Quaternary glaciations. Another difficulty of such a comparison is the absence of reliable data on the average latitudinal boundaries of the ice cover during various ice ages. It is therefore possible to compare only the main results of our calculations with the empirical data on the natural conditions during the ice ages, and this particularly applies to the ice cover.

The maximum calculated migration of the average ice boundary in the northern hemisphere accords well with the empirical data. For example, Lamb (1964) found that during the stage of greatest glaciation the average ice boundary in the northern hemisphere expanded as far as 57°N, which corresponds to a 15° shift of this boundary as compared to its present position. In our calculations this shift is 12°, which is close enough to Lamb's result.

It is difficult to carry out this comparison for all ice ages but it may be done for those regions of the earth for which data are available. For example, Zeuner (1959) gives information about the expansion of glaciers during various epochs in Central and Northern Europe. This spread, expressed by Zeuner in percentages of the distance travelled by the glaciers during the Mindel II epoch, are compared in Fig. 21 to the calculated variations of the boundary latitudes of the north polar ice during the Riss I, Riss II, Würm I, Würm II, and Würm III epochs. Since Zeuner gives a range of values for each of these epochs, the results are shown on the diagram in the form of line segments. As we can see from the diagram, there is a certain correlation between the quantities, which proves that it is possible to estimate the relative parameters of different glaciations by a quantitative method.

The conclusion that the development of Quaternary glaciations could be caused by astronomical factors does not exclude the effect of volcanic activity on glaciations. According to contemporary observations, fluctuations in the transparency of the stratosphere due to the changing amount of aerosol particles cause variations in the total radia-

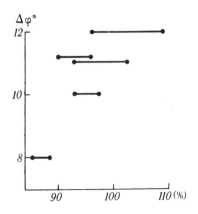

Fig. 21. Expansion of glaciers during various epochs.

tion, averaged over 10-year periods, which amount to several tenths of a percent.

Calculations carried out according to this scheme show that if such variations in the radiation had lasted for several thousand years, they could have caused a migration of the ice boundary of up to several hundred kilometers. Although it is known from geology that volcanic activity in the past varied enormously, it is at present difficult to estimate the long-term variations in the amount of atmospheric aerosol. This limits any opportunity to make a more detailed study of the relation between volcanism and the Quaternary glaciations.

We note that this agreement between the calculated distribution of the polar ice and the paleogeographic data could be reached only by taking into account the feedback between the polar ice and the atmospheric thermal regime. An interesting example of a study which ignored this feedback is given in the work of Saltzman and Vernekar (1971). They used a numerical model of the average latitudinal distribution of temperature, winds, evaporation and precipitation, based on the integration of the equations of atmospheric dynamics. By this model they calculated the variations of the temperature distribution at the surface in the northern hemisphere that took place 10 and 25 thousand years ago, as compared with that of today. All the climatic factors were assumed to be constant, with the exception of the radiation at the outer boundary of the atmosphere, which differed from the present value due to astronomical factors.

Saltzman and Vernekar came to the conclusion that during those periods, the largest temperature difference at certain latitudes for the warm half-year did not exceed 1.0°. They concluded that such minor variations were inadequate for the development of glaciations.

Since Saltzman and Vernekar did not take into account the feedback between the ice cover and the temperature field, we can expect their temperature values to be underestimates. It is interesting to compare their results to those of a similar calculation done by means of our model without feedback, i.e. considering the albedo of the earth-atmosphere system constant as the radiation varies. The results of this calculation are presented in Fig. 22, which gives the distribution of average latitudinal temperature differences for the warm half-year 10 and 25 thousands of years ago, compared to modern conditions.

It follows from the figure that the obtained results correlate quite well, and this is rather remarkable since the Saltzman-Vernekar model differs greatly from our more schematic model. The satisfactory agreement of these calculations based on different models testifies to the reliability of these models and to the possibility of making many simpli-

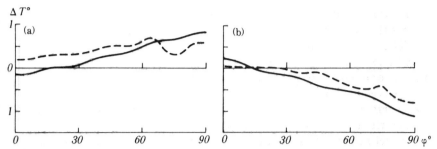

Fig. 22. Distribution of average latitudinal temperature differences for the warm half-year caused by astronomical factors. (a) 10 thousand years ago; (b) 25 thousand years ago. Solid line, calculations by means of the model described in Chapter 2; broken line, calculations by Saltzman and Vernekar.

fying assumptions in developing the semiempirical theories of the thermal regime.

One of the reasons for the agreement between the results of the thermal regime variations calculated from the above semiempirical model and those obtained using more general climatic theories presumably is the fact that the basic empirical hypothesis used in the indicated model, and expressed as (2.5), is fulfilled not only for mean annual conditions but also for different seasons, and is confirmed by observation.

It is known that the average temperature difference between the pole and the equator changes in the course of the year by several tens of degrees; this corresponds to extremely great variations in the incoming energy, producing large-scale atmospheric motions. At the same time the chief patterns of the summer and winter global atmospheric circulation in each hemisphere remain virtually unchanged, making it possible to describe the meridional heat transfer during various seasons by similar empirical relations.

Of interest from this standpoint is the work of Williams, Barry and Washington (1973), in which a numerical model of the atmospheric circulation during the last Würm glaciation was constructed.

Flohn and other authors have voiced the opinion that during the ice ages summer climatic conditions in middle and high latitudes resembled modern winter conditions. Williams, Barry and Washington arrived at a similar conclusion by physical deduction.

It follows from their work that during the ice ages the summer westward transfer in the middle latitudes of the northern hemisphere was approximately the same as during contemporary winter conditions. Under the influence of glaciation, the stable pressure systems were shifted somewhat and there was some redistribution of precipitation.

There were no drastic changes, however, in the atmospheric circulation during that period. Similar results were obtained by Alyea (1972).

Thus it is apparent that the simplified numerical models that satisfactorily describe modern climatic conditions for various seasons, may be applied to the Quaternary period as well.

Less clear, however, is the question of whether it is possible to apply the thermal regime models for average annual conditions in order to discern the mechanism by which Quaternary glaciations developed.

Fluctuations of astronomical factors do not affect the total annual radiation as heavily as the total radiation for the warm and cold half-years separately. Since the ice cover regime is controlled mainly by the thermal conditions during the warm season, it is obvious that the application of average annual data can underestimate the effect of astronomical factors on glaciation. This is confirmed by our calculations and by Sellers (1969, 1970). Sellers used his model of the atmospheric thermal regime for studying the relation between astronomical factors and the Quaternary glaciations, and came to the conclusion that the dependence was negligible. On the other hand, Berger (1973) applied Sellers' model for average annual conditions and obtained the ice boundary migration during the Quaternary period, a result quite consistent with paleogeographic data.

Calculations based on the semiempirical model described in Chapter 2 involving the feedback between the ice cover and the thermal regime have shown that in this case the influence of the radiational factors on the air temperature in high latitudes is tremendously enhanced, making it possible for the glaciations to expand over large areas. It may be assumed that if this feedback is taken into consideration, a similar result can be reached with the Saltzman-Vernekar model as well as other available models of the atmospheric thermal regime.

It follows from the foregoing that astronomical factors began to significantly influence climatic conditions only after the appearance of large polar glaciations, since it is the latter that made the Quaternary climatic conditions so sensitive to the minute fluctuations of climatic factors. During earlier epochs, before the formation of the polar ice sheets, the astronomical factors varied in a similar manner as during the Quaternary, but their effect on the climate was much smaller. The extent of this influence seems to correspond to that calculated on the basis of our model and the Saltzman-Vernekar model, as illustrated by Fig. 22.

This figure implies that fluctuations of astronomical factors in the absence of ice covers change the air temperature by at most 1°C.

Therefore, during the Pre-Quaternary times, astronomical factors could not have affected the climate to any great extent.

Thus, in order to explain the origin of Quaternary ice ages, it is important to find the mechanism of temperature decrease in the polar zones that took place during the Pliocene.

The Role of Atmospheric Carbon Dioxide. This problem is analyzed in Section 4.2, where it is established that fluctuations of carbon dioxide concentration in the atmosphere did not affect the thermal regime while it remained above 0.10%. Under those conditions the air temperature was about 5°C higher than at present.

When its concentration dropped below this level, the air temperature began to decrease, slowly at first, and then more rapidly.

Taking into account the estimate given below of change in CO_2 concentration during the Cenozoic era it appears that the lowering of temperature caused by a decrease in CO_2 concentration began in the Oligocene and accelerated during the Pliocene.

It follows from the formulas of the semiempirical theory of the thermal regime that the first polar glaciation must have developed in the southern hemisphere, where the major part of the polar zone was occupied by the Antarctic continent.

This process was accelerated by the influence of mountains with even lower surface temperatures and by the absence of ocean currents that supply an influx of heat in the north polar zone. According to the semiempirical theory, the contemporary size of this glacier provides for a 2°C drop in the global air temperature at the earth's surface. It is therefore possible that its development facilitated the origin of the ice cover in the northern hemisphere.

Thus, the development of the polar glaciations could have been induced by the drop in the carbon dioxide concentration, assuming that the continents were situated favorably for the temperature decrease at the poles.

It is interesting to ascertain whether fluctuations in the carbon dioxide concentration during the Quaternary period were adequate for the expansion and recession of the ice cover. The next section will show that the boundaries of the polar glaciation can migrate by 5–20° latitude, when the concentration changes by 0.005–0.010%, i.e. by $\frac{1}{6}-\frac{1}{3}$ of its present amount. Although such fluctuations are comparatively small, there are no indications whatsoever that they actually occurred during that time.

On the other hand, there can be no doubt that the orientation of the earth's surface relative to the sun changed periodically during the

Quaternary. Since the calculations cited above prove that the variations in the radiation regime caused by this effect are sufficient for the development of glaciations, we need not count the fluctuations of carbon dioxide concentration among the phenomena causing the migration of the ice cover boundaries.

The Role of Volcanic Activity. It was mentioned in Chapter 1 that several authors point to volcanic activity as a major factor of climatic variations during the Quaternary.

It follows from general considerations that fluctuations in volcanic activity can affect climatic conditions in two ways: by increasing the amount of carbon dioxide in the atmosphere and by reducing atmospheric transparency.

The first of these mechanisms led to a rise in air temperature during the Quarternary, when the level of carbon dioxide was low, while the second mechanism has reduced the temperature throughout geologic time.

We can assume that during the Quaternary the second effect prevailed over the first one. If a volcanic eruption ejects into the atmosphere similar amounts of carbon dioxide and sulphurous gas, their effect on the thermal regime is quite different. A comparatively small amount of sulphurous gas (say, several million tons) can reduce the atmosphere's transparency considerably by producing aerosol particles in the lower stratosphere, and this can cause a sensible drop of air temperature at the earth's surface (see Chapters 3 and 7). At the same time an equal or even a thousandfold addition of carbon dioxide changes the mass of atmospheric carbon dioxide very little and, consequently, cannot possibly affect the climate.

The conclusion is confirmed by the noticeable increase in the mean global temperature during a quiet volcanic period (1920s–1930s).

In accordance with the data given in Section 4.2, we may conclude that in the course of the last hundreds of millions of years there was a tendency toward a reduction in the amount of carbon dioxide in the atmosphere, but there is no proof of any noticeable increase of its concentration during the Quaternary.

Since the volcanic activity in the Quaternary period led to a decrease of air temperature, it is important to estimate its possible value.

In the next section we go on to analyze the mechanism of drastic short-term air temperature changes over the planet caused by the statistical regularities of the simultaneous eruption of a number of volcanoes. In the course of hundreds of thousands of years these events could have resulted in numerous air temperature decreases amounting to several degrees over several decades. It might be thought that such temperature changes had no considerable influence upon the development of glaciation.

It is possible that the fluctuation in the number and power of volcanic eruptions for time intervals of hundreds and thousands of years could have played a great part in the formation of climatic conditions during the Quaternary.

From Chapter 3 it follows that the rise in global temperature in the first half of the twentieth century was limited to about 0.5°C chiefly because the period of increased atmospheric transparency was so brief. Had this period lasted several centuries instead of several decades, the polar sea ice would have melted considerably, and this would have meant an additional considerable temperature increase. A similar result could be reached by a long-term reduction of the air transparency which attenuates the total radiation flux by several tenths of a per cent, i.e. by an amount corresponding to the current climatic fluctuation. Such a reduction would be sufficient to produce a migration of the polar ice boundary into lower latitudes by hundreds of kilometers, which is on the same scale as the ice expansion during the major glaciations.

Thus, if changes in volcanic activity which occurred in the course of the last century had continued for a sufficiently long time, say, thousands or tens of thousands of years, they could have brought about the ice ages.

Unfortunately, the available material on the frequency of volcanic eruptions is not sufficient for calculating these fluctuations over prolonged periods of time. Statistical estimates based on the recurrence of eruptions as a random process are more reliable for brief periods of time, when the number of eruptions remains considerably lower than the number of active volcanoes. From materials on epochs of higher volcanic activity that undoubtedly took place, it is difficult to estimate the transparency changes in the atmosphere and the duration of these changes in the course of the given epochs.

In estimating the role of volcanic activity in the development of Quaternary glaciations, therefore, the deduction that changes in astronomical factors are sufficient for the development and melting of glaciations is highly significant.

From this it follows that the volcanic activity has not been the major factor in the formation of the ice sheets, although its fluctuations might have had some effect on the Quaternary climatic conditions. This effect was probably not as strong as that of the astronomical factors and apparently manifested itself mainly during briefer time periods.

It is interesting to analyze in this connection the mechanism of climatic variations during the last several thousand years, a short description of which has been given in Chapter 1.

These fluctuations lasted only about several hundreds or thousands of years, thus excluding the possibility of any relation to the astronomical climatic factors which are of considerably greater duration.

Auer (1956) and other scientists have established the following epochs of increased volcanic activity: ca. 7000 B.C.; 3500–3000 B.C.; 500–200 B.C., and A.D. 1500–1900. Lamb (1970) contends that the volcanic activity could have affected the climatic conditions during these epochs but it was not the only cause of these climatic variations.

The climatic fluctuations during the Holocene were characterized by comparatively small changes in the air temperature. They could have been produced by fluctuations in the atmospheric transparency due to volcanic activity. They could also have been caused by auto-oscillatory processes occurring in the ocean-polar ice-atmosphere system.

Causes of Climatic Variations. The specific climatic conditions of the Quaternary period were apparently due to a drop in the carbon dioxide concentration in the atmosphere and to continental drift and orogeny, which will be analyzed in the next section. This has led to a partial isolation of the Arctic Ocean and to the shift of the Antarctic continent to the polar zone of the southern hemisphere.

The Quaternary period was preceded by a climatic evolution of long duration leading to the formation of distinctive thermal zones. This process was caused by changes on the earth's surface and found expression mainly in the air temperature decrease in the middle and high latitudes.

During the Pliocene climatic conditions began to be affected by the lower concentration of carbon dioxide in the atmosphere, which led to a drop in the global air temperature of 2–3°C (and 3–5°C in high latitudes). After this the polar ice covers began to develop and this led to a further reduction in the mean global temperature, especially in high latitudes.

The existence of polar ice caps sharply increased the sensitivity of the thermal regime to small fluctuations in climatic factors. Thus, enormous boundary migrations of the snow and ice covers on land and sea became possible due to the changes in the position of the earth's surface with respect to the sun, which previously had not exerted a noticeable effect on the climate.

The continued decrease of the carbon dioxide concentration in the atmosphere facilitated the intensification of later glaciations as compared to earlier ones, although the scale of glaciation depended primarily on the combination of astronomical factors.

It seems that astronomical factors exerted the greatest influence on climatic variations during the Quaternary, though it is very probable that the changes in volcanic activity could also play an important role in climatic variations during time intervals of several hundreds or thousands of years.

4.2 Pre-Quaternary Variations

As pointed out in Chapter 1, the farther back we go into the ages, the less amount of evidence do we find on climatic conditions, and the more difficult it becomes to interpret these data.

The most reliable, though schematic, information on the climates of the past stems from the very fact of the continuous existence of living organisms on our planet. The ecology of organisms of the past may of course have differed greatly from that of contemporary related forms. Nevertheless it is highly improbable that they could have subsisted outside the limits of a comparatively narrow temperature range, say, from 0 to 50°C, within which the overwhelming majority of present-day animals and plants live.

It may therefore be assumed that the surface temperatures, as well as that of the lower atmospheric and upper water layers, did not exceed these limits over the greater part of the earth during the past 2–3 billion years, i.e. since living organisms first came into being.

The actual fluctuations of the mean surface temperature over long time intervals were smaller than this temperature range, and in most cases, did not exceed several degrees over millions of years.

Hence it is extremely difficult to study past variations in the earth's thermal regime on the basis of empirical data, because the errors in temperature determined either by means of radioisotopic analysis of organic remains or by any other known method, are generally not less than several degrees.

Another difficulty in studying the climates of the remote past mentioned in Chapter 1 concerns the location of different regions relative to the poles, caused by continental drift and possibly the motion of the poles themselves.

We shall therefore limit ourselves to an analysis of the secular variations of air temperature that occurred during the second half of the Mesozoic era and the Tertiary period. We will be using paleotemperature determinations obtained by Emiliani (1966) by means of radioisotopic analysis.

Curve 1 in Fig. 23 gives the secular distribution of the mean temperature of the ocean surface level in middle latitudes, according to Emiliani. We see from this figure that the air temperature near the earth's surface during the last 130 million years had a tendency to decrease. This tendency was interrupted several times by increases in

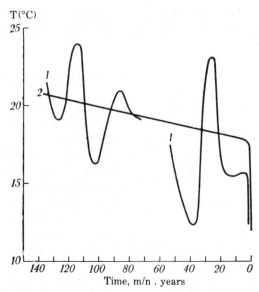

Fig. 23. Secular temperature variations in the middle latitudes of northern hemisphere. (1) Observational data; (2) calculated curve.

the air temperature, which, however, gave way to cooling and did not change the general trend of climatic evolution.

In order to explain these climatic fluctuations, it is necessary to analyze the influence of the atmosphere's composition and the structure of the earth's surface on the thermal regime.

Atmospheric Carbon Dioxide. In studying the Pre-Quaternary climatic conditions it is important to deal with the effect of atmospheric carbon dioxide decrease on the thermal regime.

Carbon dioxide is one of the comparatively small constituents of the atmosphere. At the present time the atmosphere contains about $2.3 \cdot 10^{12}$ t of carbon dioxide which amounts to 0.032% of the entire volume of the atmosphere.

A much larger amount of carbon dioxide is present in the hydrosphere, where $130 \cdot 10^{12}$ t of carbon dioxide is dissolved, primarily in the oceans. A constant exchange of carbon dioxide goes on between the atmosphere and the hydrosphere by means of molecular and turbulent diffusion.

Measurements of carbon dioxide concentration in the atmosphere show that it changes very little with location and also with altitude within the troposphere. The relatively constant amount of carbon dioxide in the atmosphere, as compared to water vapor, is explained by the more stable nature of the sources and sinks of carbon dioxide on the earth's surface and by its low sensitivity to temperature variations. Observational data show that the carbon dioxide concentration increases slightly in the equatorial region and drops by approximately 0.005% in high latitudes. This effect is explained by the higher solubility of carbon dioxide in the cold ocean waters in high latitudes, as compared to warm tropical waters. As a result, the atmosphere in high latitudes loses its carbon dioxide, which is then dissolved in the oceans. Its surplus is transported by the deep, cold currents to the lower latitudes where it is released back into the atmosphere. The flux of carbon dioxide between the equator and the north pole produced by this mechanism is about $2 \cdot 10^{10}$ t/yr (Bolin and Keeling, 1963).

A continuous carbon dioxide exchange exists between the atmosphere and the hydrosphere, on the one hand, and living organisms and the lithosphere, on the other. This cycle plays an extremely important part in the maintenance of life on our planet.

The main ingredients of this cycle are determined by biological processes. Carbon dioxide is absorbed by autotrophic plants during photosynthesis, which is the source of the direct or indirect energy of the overwhelming majority of living organisms. In the course of their biologic activity (particularly, during the breathing process) carbon dioxide is released and returned to the atmosphere.

A certain amount of carbon dioxide is ejected into the atmosphere from the deep layers of the earth's crust during volcanic eruptions, from mineral water sources, etc. Carbon dioxide is consumed by silicate rocks during weathering and in the process of formation of various carbonaceous combinations.

Thus, two cycles, biological and geological, are involved in the balance of atmospheric carbon dioxide, and each of these has its sources and sinks of carbon dioxide. Available data show that the biological components of the annual carbon dioxide cycle considerably exceed the geological components of this process.

The annual carbon dioxide expenditure for photosynthesis at the present time is about 10^{11} t. The same amount of carbon dioxide is produced in the process of breathing and the decomposition of living organisms. A much smaller amount of carbon dioxide, about 10^8 t, comes from volcanic eruptions. An amount of carbon dioxide of the same order of magnitude is used up in various geological processes (Plass, 1956; Müller, 1960; Lieth, 1963).

Living organisms, products of their metabolic activity, and the lithosphere contain large amounts of carbon drawn from the atmosphere and hydrosphere in the form of carbon dioxide. The amount of carbon in living organisms apparently corresponds to about 10^{12} t of carbon dioxide. The estimates of different researchers vary considerably as to this value (Takahashi, 1967; Man's Impact on the Global Environment, 1970).

The amount of carbon in the lithosphere corresponds to about $2 \cdot 10^{17}$ t of carbon dioxide, the major part of which is trapped in carbonaceous rocks (Vinogradov, 1972).

Although the estimates given of both the components of the carbon dioxide balance, and of the carbon stored in various natural environments, are very approximate, some conclusions may be drawn about the rate of the carbon dioxide cycle.

For example, the living organisms have a comparatively low "carbon dioxide capacity"; the amount of carbon in them is renewed in a rather brief period of about 10 years, while in the atmosphere this process takes about 20 years.

The period during which carbon has been accumulating in the lithosphere is very long and is commensurate with the period of existence of the biosphere.

There are grounds for believing that the carbon dioxide concentration in the atmosphere during the geological past differed greatly from the present. It follows therefore that the carbon dioxide balance in the atmosphere has not always been equal to zero.

However, this difference between the gain and loss of carbon dioxide comprises a very small fraction of the absolute values of the main components contributing to the carbon dioxide budget of the atmosphere. For instance, if we assume that over the last million years the carbon dioxide concentration in the atmosphere dropped to one half, its annual variation would correspond to only 0.002% of the annual amount of photosynthesis.

Geochemical studies indicate that during the last part of the geological history of the earth, the carbon dioxide concentration in the atmosphere has tended to decrease.

According to Vinogradov (1957), the atmospheric carbon dioxide originated together with other gases in the course of the zonal molting of the earth's mantle, heated in the process of radioactive decay of the elements. During the Proterozoic era, the atmosphere was anaerobic and contained a large amount of carbon dioxide. By the end of that time the composition of the atmosphere began to change under the influence of photosynthesis, the amount of oxygen gradually increasing and the atmosphere acquiring an oxidizing property. The increase in oxygen content was accompanied by a decrease of carbon dioxide.

Ronov (1959, 1964, 1972) studied the regularities of changing the amount of carbon dioxide in the atmosphere during the geological past and established that throughout the end of the Paleozoic and Mesozoic eras the carbon dioxide concentration dropped gradually, chiefly as a result of its consumption in the formation of carbonaceous rocks. The amount of carbon dioxide in the atmosphere increased somewhat in the period of higher volcanic activity.

Estimates of the atmospheric carbon dioxide during various geological periods are highly approximate. For example, Hoffman (1951) suggests that its concentration during the Permian period (about 250 million years ago), was 7.5%. On the basis of this figure we come to the conclusion that during the Mesozoic and Cenozoic eras, the average rate of its decrease came to 0.03% per million years.

There are also other opinions on the rate of carbon dioxide changes during the geological past. For example, Rutten (1971) contends that its maximum concentration during the Phanerozoic exceeded its modern level by not more than a factor of ten. This agrees with the assumption that the main part of the free carbon dioxide was removed from the atmosphere during the Proterozoic era, after which the rate of its decrease dropped drastically.

In the study carried out by the author jointly with Ronov and Yanshin, using data on the carbon content in sediments, the change in atmospheric CO_2 mass during the last 570M yr was computed. Fig. 29 (curve CO_2) shows the results of this calculation. From this figure it is inferred that during this time interval the CO_2 concentration changed approximately by an order of magnitude. This conclusion fits the view point of Rutten. From the end of the Mesozoic era the tendency for CO_2 mass to decrease prevailed, resulting in lowering its concentration from a few tenths to three hundredths per cent.

Let us consider the influence exercised by the carbon dioxide concentration on living things.

The vegetative cover of the land consists mainly of autotrophic plants producing organic matter from atmospheric carbon dioxide, water, and minerals from the soil. Together with the atmospheric carbon dioxide, plants can utilize the carbon dioxide from the soil; its amount, however, is negligible as compared to that in the atmosphere. The autotrophic plants of the hydrosphere utilize the carbon dioxide dissolved in the water for their photosynthesis.

It has long been established that under sufficient insolation and other favorable conditions, the rate of photosynthesis is strongly limited by the low concentration of carbon dioxide in the atmosphere and water reservoirs.

It has been proved by numerous experiments that an increase in carbon dioxide concentration facilitates the increase of the photosynthetic rate, other conditions being equal. For most plants this rate is higher when the carbon dioxide concentration increases up to several tenths of one per cent. After this the photosynthesis rate stops rising, and when the concentration exceeds several per cent, the photosynthesis begins to drop.

The considerable enhancement of photosynthesis under increased carbon dioxide concentration is often interpreted as evidence of the adaptability of autotrophic plants in the course of their evolution to higher concentrations in the past.

Since the geochemical data indicates a gradual decrease in the amount of atmospheric carbon dioxide during the last hundred of millions of years, this assumption can be considered plausible.

The question arises in this connection as to whether, during the time of the existence of autotrophic vegetation, the carbon dioxide concentration could exceed the value at which the photosynthesis rate begins to drop.

It is rather difficult to answer this question because the dependence of photosynthesis on the amount of carbon dioxide for the plants of the remote past does not necessarily coincide with a similar dependence that exists for present-day plants.

Let us note that the variation in the amount of carbon dioxide in the atmosphere has an effect on the total mass of living organisms on our planet. The major portion of this biomass consists of autotrophic plants, the amount of which is controlled by their photosynthesis. This dependence has a very simple explanation. The consumption of organic matter by autotrophic plants through breathing, through the dying of certain parts of plants, etc. can be considered proportional to plant mass. Inasmuch as the total amount of photosynthesis is equal to the expenditure of organic matter and the rate of photosynthesis is approximately proportional to the concentration of carbon dioxide, we find that the mass of vegetation is also proportional to the concentration of carbon dioxide.

We may assume that in some cases the dependence of the vegetative mass on the carbon dioxide concentration is somewhat weaker, since the rate of photosynthesis can increase with the surface area of the plant. However, for a compact vegetation cover with an optimal structure providing the maximum photosynthesis, the dependence of the total photosynthesis on the size of the photosynthesizing surface is not great (Budyko, 1971).

Thus the mass of autotrophic vegetation depends greatly on the carbon dioxide concentration and sometimes the mass is directly proportional to the concentration.

Obviously, this conclusion pertains mainly to those cases in which the amount of water, mineral nourishment and other factors are in sufficient supply, that is, they do not thwart the photosynthesis process.

It is possible, however, that even if there were unfavorable factors, some relation would exist between the vegetation mass and carbon dioxide concentration, one that would be particularly noticeable at low concentrations.

From this it follows that during the geological past, when the carbon dioxide concentration was considerably higher than it is now, the mass of plant cover on land was probably greater than at present. This refers chiefly to vegetation in zones with sufficient moisture. The total amount of the biomass of autotrophic vegetation in the ocean during the time of high carbon dioxide content could also be higher than its present level.

Thus, the process of the carbon dioxide decrease in the atmosphere and hydrosphere was probably accompanied by a gradual decrease in the mass of autotrophic plants and consequently in the mass of all living organisms on our planet.

This law implies that for every plant of a certain size there exists a lower boun ·ary of carbon dioxide concentration below which the given plant cannot go on living. Under these circumstances, autotrophic microorganisms last longer than any others, due to the fact that the ratio of their mass to the area of the photosynthesizing surface is minimal.

The question arises as to the limit of the diminishing global biomass if the tendency of carbon dioxide to decrease continues in the future. As noted above, over the last 100 million years CO_2 concentration decreased by approximately 0.2%. Since its present value is only about 0.03%, it seems obvious that autotrophic plants will disappear within a few million years, which is a short period compared to the duration of their existence.

There are, however, insufficient grounds for this forecast. During the most recent times the decrease in the amount of carbon dioxide ceased and gave way to an increase due to the economic activity of man. There are reasons to believe that, had this process of carbon dioxide decrease continued further, there would have been a drastic change in the climatic conditions that would have made the continuation of life impossible on this planet, and this would have happened long before the dying-out of autotrophic vegetation caused by an insufficiency of carbon dioxide.

Climatic Effects of Atmospheric Carbon Dioxide. Let us consider the dependence of air temperature on the carbon dioxide concentration in the atmosphere and on the albedo of the earth-atmosphere system (Budyko, 1973). For this purpose we shall make use of the semiempirical model of the atmospheric thermal regime described in Chapter 2.

Let us assume that in the absence of feedback between the air temperature and the polar ice, the temperature at the earth's surface varies with carbon dioxide concentration, in accordance with the dependence established by Rakipova and Vishnyakova (1973).

To estimate the effect of carbon dioxide on the thermal regime by considering the feedback, we shall use the relation

$$I_s' = \gamma I_s, \qquad (4.2)$$

where I_s is the long-wave radiation lost into outer space for the present value of carbon dioxide concentration in the atmosphere; I_s' the outgoing radiation for various values of carbon dioxide concentration, γ a dimensionless coefficient depending on the carbon dioxide concentration. This relation can be considered sufficiently exact when the coefficient γ is close to unity.

By using the empirical relation given in Chapter 2 for determining I_s we find

$$I_s' = \gamma[a + bT - (a_1 + b_1 T)n], \qquad (4.3)$$

where T is the air temperature at the earth's surface in °C, n the cloudiness in fractions of unity; a, b, a_1, b_1 the dimension factors.

Considering that for the earth as a whole the amount of absorbed radiation is equal to the outgoing radiation we find the relation

$$Q_{sp}(1 - \alpha_{sp}) = \gamma[a + bT_p - (a_1 + b_1 T_p)n_p], \qquad (4.4)$$

where Q_{sp} is the solar radiation at the outer boundary of the atmosphere, α_{sp} the albedo of the earth-atmosphere system, T_p the mean planetary temperature at the earth's surface, n_p the average cloudiness. For determining the average planetary temperature we obtain the formula

$$T_p = \frac{Q_{sp}(1 - \alpha_{sp}) - \gamma a + \gamma a_1 n_p}{\gamma(b - b_1 n_p)}. \qquad (4.5)$$

Assuming that the coefficient γ varies only slightly with latitude, and applying the semiempirical theory of the atmosphere's thermal regime for determining the mean annual temperature of various latitude zones, we obtain the formula

$$T = \frac{Q_s(1 - \alpha_s) - \gamma a + \gamma a_1 n + \beta T_p}{\beta + \gamma b - \gamma b_1 n}, \qquad (4.6)$$

where Q_s, α_s, n, and T refer to the latitude zone under consideration; β is a dimension factor describing the intensity of the meridional heat exchange.

The γ coefficient for various concentrations of carbon dioxide can be

4 CLIMATES OF THE PAST

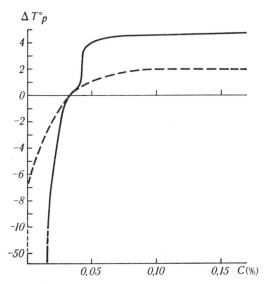

Fig. 24. Effect of carbon dioxide on air temperature. Solid line, albedo depending on air temperature; broken line, constant albedo.

found from formula (4.4) by using it together with the relation established by Rakipova and Vishnyakova and presented in Fig. 24 by curve 1. The determined values of this coefficient are given in Table 14.

By taking into account these values and the relation between the boundaries of the polar ice cover and the albedo, on the one hand, and the air temperature, on the other, we can calculate the variations of the mean planetary temperature for the changing concentration of carbon dioxide with an existing feedback between the polar ice and the thermal regime. The result of this calculation is illustrated in Fig. 24 by curve 2.

As we see from Fig. 24, fluctuations in the surface area of the polar ice greatly enhance the effect of the carbon dioxide fluctuations on the mean planetary temperature. When the carbon dioxide concentration increases 30% above its present level, the temperature rises significantly, and when the amount of carbon dioxide exceeds 0.05%, the temperature rises several degrees above its present level.

When the amount of carbon dioxide drops, so does the average

Table 14
Dependence of Coefficient γ on Carbon Dioxide Concentration

C (%)	0.000	0.005	0.010	0.020	0.032	0.050	0.070	0.100
γ	1.038	1.030	1.030	1.011	1.000	0.994	0.990	0.988

planetary temperature, and when the carbon dioxide concentration is equal to about half of its present amount, the average temperature at the earth's surface drops by several tens of degrees which corresponds to conditions of total glaciation of the earth.

The dependence of the polar ice boundary in the northern hemisphere on the amount of carbon dioxide calculated according to this scheme is presented in Fig. 25. The increase of the carbon dioxide concentration by 30% above its present level (designated by a vertical line) appears to be sufficient for complete melting of the polar ice, while a reduction to half its present value leads, as pointed out above, to the total glaciation of our planet.

It should be noted that the general relation between the average air temperature and the polar ice boundary, on the one hand, and the carbon dioxide concentration, on the other, is more complicated than the curves shown in Figs. 24 and 25. This relation is similar to the dependence of these parameters on the solar constant which is ambiguous, as proved in Chapter 2.

Since in this case we are concerned solely with the climatic effects of variations in the carbon dioxide concentration from its present level, the pertinent relations become simplified and obtain the form given above.

When analyzing the relations presented in Figs. 24 and 25, it is necessary to remember that they characterize stationary conditions of the atmosphere-ocean-polar ice system that can be reached only over long periods, such as several thousand years. If the changes in the carbon dioxide concentration occur at a much higher rate (as, for instance, is the case at the present time), then the variations of the thermal regime and the polar ice boundary will be considerably smaller than those determined by the relations presented in Figs. 24 and 25.

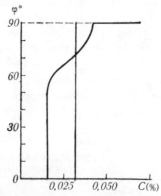

Fig. 25. Dependence of polar ice boundary in northern hemisphere on carbon dioxide concentration.

By means of formulas (4.5) and (4.6), we can calculate the distribution of mean latitudinal temperatures at the earth's surface for two cases: 1) an ice-free regime corresponding to a carbon dioxide concentration of 0.042%, and 2) the total glaciation of the earth corresponding to a concentration of 0.015%.

In the first instance, calculations were carried out under the assumption that the albedo of the earth-atmosphere system in high latitudes is equal to 0.40 in the absence of ice. The effect on the thermal regime of the possible changes in cloudiness was neglected.

In the second instance, it was assumed that the albedo at all latitudes was equal to 0.64 and that the effect of cloudiness was negligible.

The results of calculations for the northern hemisphere are presented in Fig. 26. We may see from this figure that when the concentration of carbon dioxide is equal to 0.042% (curve 1), the mean annual temperature in low latitudes rises approximately 2°C and in high latitudes as much as 10°C compared to present conditions (curve 2).

In the case of total glaciation (white earth) the mean annual temperature drops approximately 70°C at the equator and 45°C at the pole (curve 3). It must be noted that these temperature values for the case of total glaciation differ somewhat from the values obtained in Chapter 2. This discrepancy is due to the fact that in the present

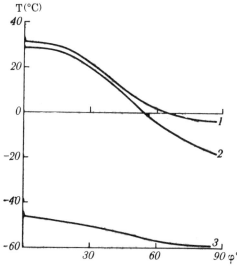

Fig. 26. Distribution of mean latitudinal temperature as a function of carbon dioxide concentration. (1) Ice-free regime; (2) present conditions; (3) total glaciation.

calculation we neglected the effect of cloudiness on the thermal regime at very low air temperatures, and we also assumed a different value for the albedo of the white earth. Since the mean air temperature at the earth's surface in the case of total glaciation is far below zero under various assumptions about the values of albedo and cloudiness, it can be deduced that the white earth regime is highly stable.

Due to the enormous effect of the feedback between the temperature of the atmosphere and the polar ice, a comparatively small variation in the carbon dioxide concentration (from 0.015 to 0.042%) changes the mean annual temperatures of the atmosphere at various latitude zones by several tens of degrees.

The precision of our calculation is limited by the fact that in our model of the thermal regime we have used several simplifying assumptions. This is why our estimates of the effect of carbon dioxide concentration on the thermal regime are very rough and must be checked and made more precise in future researches.

Thus, Manabe's and Wetherald's (1975) results are of interest. They calculated the changes in the atmosphere's thermal regime for a doubling of the amount of carbon dioxide. For this purpose a three-dimensional model of the general climatic theory was used that takes into account the effect of water phase transitions and feedback between the thermal regime and the snow-and-ice cover on the air temperature.

The obtained variations of the mean latitudinal air temperature at the earth's surface obtained by Manabe and Wetherald are presented in Fig. 27 (curve M). For comparison the same figure gives the variations of the air temperature calculated according to the semiempirical model of Chapter 2, for a case when the local temperature effect of carbon

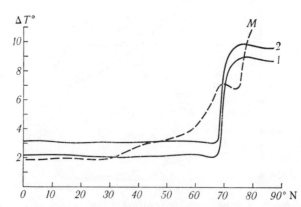

Fig. 27. Effect of carbon dioxide concentration on mean latitudinal air temperature.

dioxide is taken into account according to the Manabe-Wetherald (1967) scheme (curve 1), and to the scheme suggested by Rakipova and Vishnyakova (curve 2).

The figure shows that temperature variations at different latitudes determined by completely different models are in agreement.

It was mentioned earlier that there has been a long-term tendency for the carbon dioxide concentration in the atmosphere to decrease. It may be assumed that the fluctuations in the amount of free carbon dioxide in the atmosphere and hydrosphere over protracted time periods cannot depend on the increase or decrease of the total mass of living organisms. For example, if we assume that this mass is equal to the values given above then its variation by 100% can be compensated for by the inflow of carbon dioxide from the lithosphere (or by its expenditure on geological processes) in about 10,000 years. If we take the previous amount of biomass to be 10 times as high, then the relative time estimate will increase only to 100,000 years, which is still small according to the time-scale of geological processes.

It is obvious that the change in the mass of living organisms has an insignificant influence on the balance of carbon dioxide, as compared with that of the gas exchange between the atmosphere and the lithosphere during long time intervals commensurate with geological periods.

Unlike living organisms, the lithosphere can accumulate a very large amount of carbon and can also release a large amount of carbon dioxide over a long period of time. Therefore the geological components of the carbon dioxide balance, which are much smaller than its biological components, play the major part in the fluctuations of the amount of free carbon dioxide.

In contrast to the biological cycle of carbon dioxide, where the gain and loss are practically equal due to the comparatively small mass of living organisms, the gain and loss of carbon dioxide in the geological cycle are poorly related. Therefore there are cases possible in which the input of carbon dioxide from the lithosphere greatly exceeds its absorption by the lithosphere, and vice versa.

Considering that, under modern conditions, there is a difference of about 10^8 t per year between the loss of carbon dioxide due to geological processes, and the input of carbon dioxide from the lithosphere, we find that the amount of free carbon dioxide in the atmosphere and hydrosphere caused by this difference changes at the rate of 10^{14} t in a million years.

If we assume that the amount of atmospheric carbon dioxide is equal to a few per cent of that in the hydrosphere, we find that the amount of atmospheric carbon dioxide probably decreases by 10^{12} t, or several hundreths of a per cent in a million years. This estimation is not at variance with the above-given rate of decrease in the concentration of atmospheric carbon dioxide over the last million years.

Thus geological processes can cause variations in the atmospheric carbon dioxide concentration. It is difficult to carry out a quantitative analysis of the effect of lithospheric processes on the carbon dioxide balance because the available data on the carbon balance in the lithosphere is far from precise. In fact, from present-day estimates, we can determine only the order of magnitude of pertinent components in the carbon dioxide balance.

Although we know very little about the changes in the carbon dioxide concentration in the geological past, we can still give a comparatively reliable estimate of the influence of carbon dioxide on the thermal regime in the atmosphere during the Pre-Quaternary time.

From the aforesaid data it follows that when the carbon dioxide concentration exceeds 0.10%, the mean global air temperature increases about 5°C over its present value. The carbon dioxide concentration apparently reached this level by the second half of the Tertiary period, and therefore, during the Mesozoic era and part of the Tertiary period, the mean global air temperature at the earth's surface was 5° higher, while the fluctuation in the carbon dioxide concentration during the Pre-Quaternary time, when its mean level was rather high, had no effects on climatic conditions.

The secular temperature variations in the Pre-Quaternary, presented in Fig. 23, demonstrate that, although the temperature then was appreciably higher than at present, it was not constant. This indicates that there were other factors then which exercised an effect on the climate.

Climatic Effects of the Structure of the Earth's Surface. In the course of the second half of the Mesozoic era and the first part of the Tertiary period (the Paleogene), the mean continental level was comparatively low, so that a considerable part of the continental platform was covered by shallow seas. It was during that time that the various continental regions repeatedly rose and fell. This, however, did not prevent the continents from being criss-crossed by wider or narrower straits and seas.

During the latter part of the Tertiary period (the Neogene), intensive tectonic activity led to the elevation of continents and to a gradual disappearance of many seas. For example, a vast body of water on the territory of what is now western Siberia, that formerly joined the tropical oceans to the polar basin, then ceased to exist. Thus the Arctic

Ocean became a more or less isolated body of water which merges only with the Atlantic Ocean and is connected with the Pacific only by the narrow Bering Strait.

It is possible that the movement of the Antarctic continent to higher latitudes was completed during the Tertiary period and that it has since then been situated concentrically relative to the south pole.

Such radical changes in the structure of the earth's surface must have had a great effect on the meridional heat exchange in the ocean.

In Chapter 2 it was asserted that the present-day meridional heat transfer in the ocean comprises approximately one-half of that in the atmosphere. It is probable that the increase in the surface area of bodies of water connecting high and low latitudes leads to an increase in heat transfer. The relation between the meridional heat exchange and the distribution of mean latitudinal temperatures can be found from the semiempirical theory of the atmospheric thermal regime.

The result of such calculations for the northern hemisphere is given in Fig. 28 where curve 1 corresponds to the distribution of the mean annual temperature under ice-free conditions in high latitudes for the present ratio of meridional heat transfer in the atmosphere and in the hydrosphere. Curve 2 gives the temperature distribution when these transfers are equal, and curve 3 corresponds to a case when the transfer in the hydrosphere is twice as large as that in the atmosphere.

In the formulas of the semiempirical theory, the ratio of the transfers is determined by the parameter β_1.

Figure 28 implies that the fluctuations in the meridional heat exchange do not affect the average planetary temperature. Accordingly, the air temperature at latitude 35° remains constant for various conditions of meridional heat exchange.

Figure 28 also demonstrates that the relative increase of the meridional heat exchange in the oceans leads to a small decrease in the mean air temperature near the equator, and to a temperature increase in the high and some of the middle latitudes.

We may assume that the decrease of the meridional heat flow caused by the elevation of continents played a certain role in the temperature decrease in middle and high latitudes during the past 100–150 million years.

Assuming that at the beginning of this period the meridional heat transfer in the oceans was considerably higher than in the atmosphere but was close to its current value at the end of that period, and taking into account the drop in the global temperature at the end of the Tertiary period due to the decrease of carbon dioxide concentration, we obtain the secular temperature variations in middle latitudes corre-

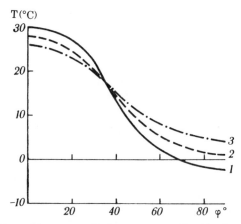

Fig. 28. Effect of heat transfer in oceans on distribution of mean latitudinal air temperature. (1) Ice-free regime; (2) meridional heat transfers in atmosphere and hydrosphere are equal; (3) heat transfer in hydrosphere is twice as large as in atmosphere.

sponding to curve 2 in Fig. 23. This curve agrees reasonably well with the empirical curve 1, though they do not coincide.

It is interesting to analyze the reasons of temperature fluctuations (see curve 1 in Fig. 23) during time intervals of tens of millions of years. Owing to the differences of opinion in the interpretation of data on paleotemperatures of the Mesozoic era and the Tertiary period, the reliability of these temperature variations and their global character is not entirely clear (Teis and Naidin, 1973). Probably these temperature fluctuations were connected with local variations in the circulation processes of the atmosphere and ocean.

A possible effect of the earth's surface structure on climate is the development of continental glaciations. At certain heights in the mountains, a constant snow cover as well as glaciation often form depending on the air temperature at sea level and on precipitation.

The lack of any traces of mountain glaciations during the Mesozoic era and the greater part of the Tertiary period is an indication not only of a warm climate but also of the low level of the continents.

From this viewpoint the nature of the last paleoglaciation – the Permo-Carboniferous – that lasted tens of millions of years, is of special interest.

As we pointed out in Chapter 1, there are traces of this glaciation in several continental regions in tropical latitudes. Paleogeographic data show that the vast regions of cold climate that appeared during the Quaternary period did not yet exist during this glaciation.

According to one hypothesis, the Permo-Carboniferous glaciation covered a continent that was then situated in high latitudes of the southern hemisphere and which later broke up.

At the same time, simple calculations indicate that due to their high albedo, large glaciations could have existed at any latitude (Budyko, 1961). This possibility follows particularly from the fact that the absorbed radiation in the region of a vast glaciation near the equator can be equal to the absorbed radiation at the pole under ice-free conditions.

If we take the albedo of the earth-atmosphere system for the ice-free polar basin as equal to 0.30, we find that the mean annual amount of absorbed radiation at a latitude of 80° is 7.7 $kcal/cm^2 \cdot month$. The same amount of radiation will be absorbed at the equator if the albedo is 0.70, which is a realistic figure for a large continental glaciation.

By means of semiempirical theory of climate, we can calculate the mean latitudinal temperature distribution in the case of an equatorial glaciation (Budyko, 1971). Let us assume that this glaciation is situated in a latitude zone near the equator occupying 5% of the total area of the hemisphere. If the albedo of this zone is equal to 0.70, the mean weighted value of the planetary albedo will drop by 0.023, as compared to ice-free conditions.

The temperature distribution, calculated according to formula (4.6), where $\beta = 0.40\ kcal/cm^2 \cdot month \cdot deg$, shows that it is nearly 0°C in the vicinity of the glacier. This proves that a glacier can exist on a plain at sea level in the equatorial region. Obviously, the parts of the glacier above this level will exist under negative temperatures.

It is interesting to note that the temperature at the pole in this case will remain positive, so that the development of the polar glaciation is not inevitable.

The possibility of existence of large stable glaciations at low latitudes does not contradict the hypotheses of continental drift and polar motion during these epochs. This conclusion only implies that the centers of large glaciations did not necessarily coincide with the poles.

Calculations of the average latitudinal temperature distribution, the results of which are presented above, were carried out without taking into account the global temperature increase under the influence of high carbon dioxide content in the atmosphere. By taking this into consideration, we see that if the Permo-Carboniferous glaciation had developed at low latitudes, it would have been mainly situated at an altitude of several hundred meters. As there is no evidence of any influence of this glaciation on the climate of other geographical regions, it is possible that this was a mountain glaciation.

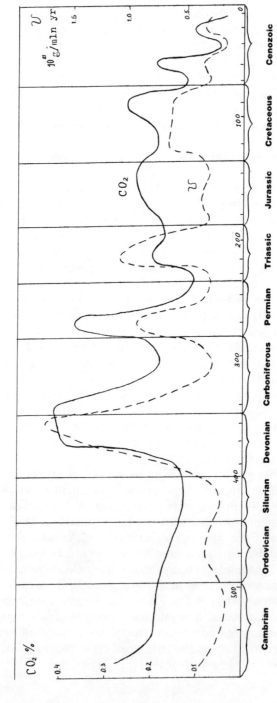

Fig. 29. Secular variations of volcanic activity and carbon dioxide content in the atmosphere. v – volcanic rocks; CO_2 – carbon dioxide content.

Climatic Effect of Volcanic Activity. The intensity of volcanic activity in the past varied to a large degree, as witnessed by the amount of eruption products contained in the layers of different geological ages. We have plotted the diagram characterizing the change in volcanic activity during the Phanerozoic eon (Fig. 29, curve V).

Figure 29 shows that the intensity of volcanic activity during different epochs underwent manifold variations. The carbon dioxide concentration changed in a more or less similar way, which is understandable as the eruptions are one of the main sources of carbon dioxide in the atmosphere. The diagram reveals a tendency toward a decrease of carbon dioxide from the second half of the Mesozoic.

As already mentioned, the climatic conditions during the main part of Phanerozoic eon did not depend on the variations of the atmospheric carbon dioxide. Therefore the only factor of volcanic activity that could affect the climate was an increase in the amount of aerosol particles in the stratosphere. As shown above, this process led to a decrease in the air temperature at the earth's surface.

In the previous section we arrived at the conclusion that the long-term climatic effects of volcanic eruptions were not great. At the same time, we believe that the short-term fluctuations could have had a perceptible influence on the atmospheric thermal regime.

Several geological studies note that the possible deviations of many natural phenomena from their normal intensity increase with time. For example, disastrous earthquakes that are hardly probable during short-term intervals become very likely over sufficiently long periods. Therefore, from general considerations, it seems natural to assume that throughout the long geological history of the earth, great anomalies in the natural processes could have taken place that have not been observed during the comparatively short time of man's existence, and particularly in the very brief period in which environmental studies have been conducted.

Let us apply this principle in estimating the volcanic activity of the past.

Since the effect of a single eruption on the temperature is comparatively low due to the limited amount of aerosol ejected into the atmosphere, it is obvious that the earth's temperature would be much more affected if a series of eruptions took place within a short time interval.

Certain difficulties arise in calculating the greatest amount of aerosol that can be ejected into the atmosphere by volcanoes over a short period of time.

An approximate estimate can be made from the table of explosive eruptions in the eighteenth–twentieth centuries, compiled by Lamb

(1969). This table also gives the amount of aerosol ejected during each eruption in relative units as compared to that of the Krakatao 1883 eruption, assumed to equal 1000 conventional units.

An analysis of Lamb's table shows that the amount of aerosol ejected into the atmosphere each decade is subject to great fluctuations, its maximum value increasing with the total duration of time, i.e. with the number of considered decades. This relation can be expressed by the simple empirical formula

$$N = a\, log_{10} \frac{T}{t}, \qquad (4.7)$$

where N is the maximum amount of dust ejected into the atmosphere during the time interval t of the period T, and the factor a is equal to 3000.

This formula is applicable to values of T much greater than t. It follows from this formula that for $T = 300$ the amount of dust ejected in a decade ($t = 10$) is equal to 4400, which is in full accord with Lamb's maximum value.

If we use this formula to extrapolate over great time intervals, we find that $N = 9000$ for $T = 10^4$, and $N = 18{,}000$ for $T = 10^7$. Thus, during a long time span, several eruptions within a single decade are possible, their combined effect from the ejected aerosol being equal to nine or even 18 Krakatao eruptions.

The reliability of this result obtained by extrapolation should not be overestimated. At the same time it must be emphasized that the above amounts of aerosol are more likely than not to be understated.

Although in Lamb's opinion the volcanic activity of the recent centuries has been rather high, there were long periods of much higher activity in the geological past. Therefore, for such periods, the aerosol amounts based on Lamb's data are not overestimated.

Let us consider the earth's temperature in the case of multiple eruptions.

If 10 eruptions comparable to the Krakatau eruption occur within a decade, the average decrease in direct radiation will be at least 10-20% (according to Chapter 3), because the rate of aerosol ejected into the atmosphere during this time is not less than its rate of settling. In this case the total radiation will drop by 2-3%, while the average temperature of the earth's surface will decrease by 3-5°.

If 20 eruptions take place within a decade, the radiation decrease will be more drastic. From atmospheric optics it follows that in this case the total radiation will drop by at least 5%, and this would lead to a temperature drop of 5–10° at the surface as this cooling spreads to both the atmosphere and the upper level of the ocean waters.

Although all these results are highly approximate, they nevertheless show that from time to time, volcanic eruptions lead to short-term but considerable temperature drops that affect the entire planet.

Under stable climatic conditions with a relatively small temperature difference between the poles and the equator that existed during the Mesozoic era and almost the entire Tertiary period, even such great short-term temperature drops could not lead to glaciation on account of the high initial temperatures in the polar latitudes.

Causes of Climatic Variations. Climatic conditions of the Mesozoic era and the Tertiary period were characterized by two main properties:

1) The mean temperature at the earth's surface was much higher than now, especially in high latitudes. Consequently, the temperature difference between the equator and the poles was smaller.

2) The air temperature had a tendency to decrease, especially in high latitudes, and this tendency grew during the last half of the Tertiary period.

These two regularities were caused by changes in two climatic factors—the amount of carbon dioxide in the atmosphere and the elevation of the continents. The higher concentration of carbon dioxide provided for an increase of 5° in the air temperature as compared with its present value. This temperature persisted through the Pliocene when, owing to a drop in the carbon dioxide concentration, the average temperature began to fall.

The low level of the continents facilitated the meridional heat exchange in the oceans, thus increasing the air temperature in middle and high latitudes.

The heat exchange in the oceans was weakened by the elevation of the continents and led to a gradual temperature drop in middle and high latitudes. The rate of cooling was accelerated during the Pliocene when the thermal regime of the atmosphere appreciably changed under the influence of CO_2 concentration variation. This cooling was most pronounced in high latitudes and made the increase of polar glaciation possible.

Although the thermal regime during the Mesozoic era and the Tertiary period was in general highly stable due to the lack of large polar glaciation, nevertheless short-term temperature drops of the air and upper water level could occur. They were caused by the frequent occurrence of a series of explosive volcanic eruptions.

5 Climate and the Evolution of Living Organisms

5.1 Ecological Systems

The Climatic Zone of Life. There are many ways in which climatic conditions determine the structure of living organisms and their spread and interaction with the environment. These conditions exert a particularly strong influence on the energy processes that sustain plants and animals.

Nearly all the primary productivity of living organisms in the contemporary epoch is determined by photosynthesis carried out by autotrophic plants (chemosynthesis supplies an insignificantly smaller quantity of organic matter than does photosynthesis).

The general pattern of energy transformations in living organisms can be represented in the form of an energy pyramid. Energy assimilated by autotrophic plants in the course of photosynthesis is at the base of this pyramid. This energy is found within the organic matter of plants. A significant part of it is consumed in the vital activity of the plants themselves, while some of it is consumed by the organisms which feed directly on the autotrophic plants.

The second tier in the energy pyramid characterizes energy influx to plants and animals which use the energy found in autotrophic plants. Usually, no more than 10% of the original energy consumed is transmitted to the living organisms which utilize the energy of autotrophic plants. These organisms also lose much energy in the course of their vital activity, in breathing, and in other functions.

The third stage of the pyramid involves the energy influx to living organisms, which exist not by means of the primary energy of autotrophic plants, but by converting energy consumed by heterotrophic organisms. This stage of the pyramid includes, in particular, carnivores.

Many living organisms are simultaneously found at the second and third stages of the energy pyramid. This is the case with man, who uses some of the energy directly created by plants and some of the energy created by animals.

5 CLIMATE AND THE EVOLUTION OF LIVING ORGANISMS

The energy pyramid reflects the close relation between animals and plants. This relation, based on the fact that plants are the only source of the energy that sustains the vital activity of all types of animals, exerts a profound influence on the body structure of animals and on their ecology.

Energy resources created by the activity of autotrophic plants define the energy base animals may exploit in different natural regions, so that the geographic location of animals and the size of their populations are functions of the energy resources.

Since the productivity of plants is conditioned to a significant degree by climatic conditions, these conditions are also a decisive factor in many zoogeographic laws.

All living organisms are highly dependent on climatic resources. The question therefore naturally arises whether organisms are able to vary climatic conditions and make them more favorable for their existence.

A comparison of the energy characteristics of climatic and biological processes persuades us that this is not so (Budyko, 1971). Such a comparison demonstrates that living organisms exploit a small part of the energy involved in climatic processes. The base of the energy pyramid characterizing the part of the influx of solar energy used by plants is much smaller than the energy involved in atmospheric processes. Typically, not more than tenths of a percent—in rare cases, several percent—of the energy of solar radiation is used in the biological cycle.

Consequently, living organisms cannot substantially vary the energy conditions of climatic processes. Animals and especially plants nevertheless affect the climate in a particular way. Such effects are basically limited to local climate and do not extend to the climate of large areas. Thus climate exerts a vast influence on the activity of animals and plants, whereas the inverse relations between living organisms and climate are comparatively weak. This feedback however, is of some importance for explaining contemporary climatic changes and for estimating possible climatic changes in the future; we will discuss such changes in detail in succeeding chapters.

The influence of climate on living organisms can be also illustrated by considering the range of conditions within which life exists on the earth.

The zone of propagation of living organisms was considered by Vernadskii, who studied the biosphere, including the atmospheric, hydrospheric, and lithospheric layers within which biological processes have developed (Vernadskii, 1926). Though Vernadskii's conception was highly fruitful, we should not forget that exaggerated conceptions as to the dimensions of the climatic zone within which life exists on earth and related conceptions as to the dimensions of the biosphere are often encountered in particular in popular literature. These conceptions are based on the idea that life pervades all regions on the surface of the globe, in the atmosphere, and in the hydrosphere.

In speaking of the zone within which living organisms are found, we should note that organisms have existed permanently in more or less significant numbers within a rather restricted space. Life is basically concentrated near the surface of the earth, though even here extensive high-latitude regions as well as a number of desert regions can be found where the amount of living organic matter per unit area is either negligible or simply zero. Living organisms are found there only by accident, and the basis for sustained life does not occur.

One further feature in the spread of living organisms is usually neglected. Life does not constitute an independent system in all areas of the globe, i.e. living organic matter cannot be created in all regions in which living organisms are found. It is well known that organic matter evolves only where autotrophic plants are found (disregarding those modes of organic matter production without any significant role in the general balance of the biomass).

However, there are extensive areas as well as a number of regions with long seasons in which the photosynthesis process is absent. Autotrophic plants cannot exist in many continental regions due to inadequate heat or moisture. The photosynthesizing activity of plants is restricted on a significant part of the ocean surface because of the absence of minerals, so that these zones are nearly barren.

Thus the energy pyramid of life clearly cannot be sustained over the entire surface of the globe. Even after billions of years of evolution, living organisms are still not able to adapt to many of the climates of our planet.

The contemporary climate is less favorable to biological processes than those climates that have prevailed over the past hundreds of millions of years.

As we noted in the preceding chapter, the decrease in the amount of carbon dioxide in the atmosphere has led to the result that its concentration is now somewhat less than that which would guarantee the highest productivity of autotrophic plants. In addition, a significant part of the earth's surface is either constantly or temporarily covered by snow and ice at the present time, and photosynthesis under these conditions is either absent or significantly diminished.

Thus the base of the energy pyramid, which represents the total mass of living organisms, is noticeably reduced in the contemporary epoch in comparison with earlier periods in the history of the earth.

5 CLIMATE AND THE EVOLUTION OF LIVING ORGANISMS

Although living organisms possess sufficient plasticity to adapt somewhat to less favorable climatic conditions, the fact that extensive zones can be found on the surface of the earth where biological processes occur slowly indicates that plants and animals are limited in their ability to adapt to changes in atmospheric conditions.

It is therefore obvious that the laws of climatic change are of enormous value for explaining the conditions under which the biosphere has developed in the past and for estimating how it will evolve in the future.

Ecological Energy Systems. Similar trophic relations unify animal and plant populations into elementary energy systems that are part of a global biological energy system. Any elementary energy system is characterized by a set of relations among the organisms comprising it and between these organisms and the environmental conditions.

Let us consider a highly schematized model of an ecological system containing four groups of living organisms:

a) autotrophic plants; b) herbivorous animals; c) predators; d) parasites utilizing the biomass of the first three groups of organisms.

Following Lotka and Volterra, we will compile balance equations for living organic matter virtually coinciding with energy balance equations.

We let B denote the biomass of a particular group of organisms per unit area occupied and P their productivity (i.e. the gain of biomass per unit time, less that consumed by their life activities); D will refer to the consumption of biomass by parasites. The concept of productivity used here corresponds to the algebraic sum of all forms of gain and loss of living energy matter, except for those forms of loss expressed by separate terms in equations (5.1)–(5.3).

The biomass balance equations for autotrophic plants have the form

$$\frac{dB_a}{dt} = P_a - b_b B_b - D_a; \tag{5.1}$$

for herbivores

$$\frac{dB_b}{dt} = P_b - b_c B_c - D_b; \tag{5.2}$$

and for predators

$$\frac{dB_c}{dt} = P_c - D_c; \tag{5.3}$$

where b_b and b_c are coefficients representing the consumption of biomass by organisms that herbivores and predators subsist on per unit mass of these groups of animals. The derivatives in the left sides of equations (5.1)–(5.3) define the rate at which biomass varies over time.

Since b_b and b_c can be considered constant to a first approximation, nine variables occur in equations (5.1)–(5.3). Consequently six additional relations are required to solve them. These relations can be compiled from empirical data for different ecological systems.

We may use the following simplified relations for a wide range of systems:

$$P_a = P_a(B_a, B_b), \qquad (5.4)$$
$$P_b = P_b(B_b, B_a, B_c), \qquad (5.5)$$
$$P_c = P_c(B_c, B_b), \qquad (5.6)$$
$$D_a = D_a(B_a), \qquad (5.7)$$
$$D_b = D_b(B_b), \qquad (5.8)$$
$$D_c = D_c(B_c). \qquad (5.9)$$

The nature of these relations has been studied in many empirical investigations.

The productivity of an autotrophic plant cover P_a, at first increases slowly as its biomass B_a grows (the surface of leaves is still small), then increases more rapidly, and again slowly when the biomass reaches a high density as a result of the decrease in photosynthesis that occurs when the lower strata of leaves are significantly shaded. The $P_a(B_a)$ dependence is represented by an S-shaped curve within a particular range of biomass increase. Productivity P_a begins to decrease at high biomass, when the additional loss for breathing is not entirely compensated by photosynthesis. The dependence $P_a(B_b)$ is determined by the fact that herbivores often increase the productivity of plants by accelerating the growth of plants, some of this increment being consumed by animals.

The productivity P_b of herbivores is usually connected to their biomass B_b by a dependence qualitatively analogous to $P_a(B_a)$, since the ratio of productivity to biomass decreases at very low and very high animal population per unit of area, chiefly due to a decrease in birth rate. Evidently, P_b depends on plant biomass B_a, increasing as the latter increases. It is affected in a similar way for a given B_b by biomass B_c of predators, who mainly eliminate sick and old animals, facilitating an increase in the birth rate.

The productivity P_c of predators varies with increasing biomass B_c in a way similar to $P_b(B_b)$, P_c increasing with increasing biomass P_b.

The use of plant and animal biomass by parasites is usually proportional to biomass, though in a number of cases this dependence is stronger than direct proportionality, since an increase in population density tends to spread parasites from one living organism to another.

5 CLIMATE AND THE EVOLUTION OF LIVING ORGANISMS

If we represent equations (5.4)–(5.9) quantitatively, we may study the laws of population dynamics for organisms within an ecological system.

A number of conclusions regarding the conditions of many ecological systems may be obtained as a result of such an analysis.

1. There is often more than one solution of these equations for time-independent values of the biomass of each group of organisms (stationary state of ecological system), which indicates that there may be different biomass values for the components of an ecological system. However, not all these solutions correspond to stable ecological systems, so that small deviations in the biomass of one group of organisms from its value for the steady state results in a return to the steady state.

2. An ecological system has some stability relative to deviations in the biomass of each of its components from the biomass in the case of the steady state. This stability is determined by the interval within which the biomass of the system components varies and within which the steady state of an ecological system is restored. Restoration of the steady state does not occur for biomass variations exceeding this interval, and one or several components of the system are destroyed.

3. The stability of ecological systems that include all these components usually exceeds the stability of systems in which some groups of animals are absent.

Since trophic interrelations (as well as nonenergy ecological relations) between the components of ecological systems confer a definite stability on these systems, such systems exhibit features of completeness (a similar point of view has been previously expressed by Elton in 1930). Ecological systems may thus exert a substantial influence on the evolutionary process.

Evolution of Ecological Systems. The population of each group of living organisms within an ecological system constantly varies.

The variation in the number n of individuals of animals of a given species over time t per unit area occupied by them is determined by the equation

$$dn/dt = \alpha n - \beta n, \qquad (5.10)$$

where α and β are the birth and death rates. If α and β are constant, letting $n = N$ at the initial moment of time, we find from (5.10) that

$$n = N e^{t(\alpha - \beta)}. \qquad (5.11)$$

When $\alpha > \beta$, the population of the animals will increase, while when $\alpha < \beta$, it will decrease.

If a population decreases, the animals in it will become extinct soon after n attains the critical value n_1 at which the low number of animals no longer insures their further reproduction by multiplication. If the population increases, an ecological crisis is reached when n attains the critical value n_2 corresponding to the population that can no longer subsist on the food resources of the given locality. In this case, destruction by the animals of their food sources is possible, which leads to a breakdown in the ecological system and threatens the existence of the species occurring in it.[1]

Bearing in mind these concepts, we find from equation (5.11) the duration for the existence of an animal population in the first and second cases:

$$t_1 = \frac{1}{\alpha - \beta} \ln \frac{n_1}{N} ; \quad t_2 = \frac{1}{\alpha - \beta} \ln \frac{n_2}{N} \qquad (5.12)$$

These equations can be rewritten in the form

$$g_1 = \frac{1}{\alpha/\beta - 1} \ln \frac{n_1}{N} ; \quad g_2 = \frac{1}{1 - \alpha/\beta} \ln \frac{N}{n_2} , \qquad (5.13)$$

where $g_1 = \beta t_1$ and $g_2 = \beta t_2$ correspond to the number of generations of animals, renewed from the initial moment of time to the moment the animals die out.

Obviously, variations in the birth and death rates are unavoidable under natural conditions, as a result of which their mean values over long intervals of time will not exactly coincide. A difference in the values of these coefficients will also occur due to variations in external conditions, whether favorable or unfavorable for a given animal species.

Equation (5.13) implies that any population may disappear comparatively rapidly, dying out once the birth and death rates do not greatly differ.

Let us bear in mind that $\ln(N/n_1)$ and $\ln(n_2/N)$ are as a rule less than 10 under natural conditions (this assumption corresponds to the possibility that the number of animals per unit area occupied by them may vary nearly by a factor of 10^9 from the minimum to the maximum values). Under these conditions a difference in the birth and death rates of only 1% will lead to the extinction of the population within a period of time less than the lifetime of several hundred generations. Since such

---

[1] A breakdown in an ecological system in a given case is to be understood as the destruction of one or more principal components of this system as a result of external factors or variations in the relations between the principal components of the system. The groups of organisms preserved following a breakdown in a previous ecological system can form a new ecological system either on their own or in combination with organisms from other systems.

an interval of time is somewhat less than the mean duration of existence of species as determined by the rate of evolution, it is obvious that the birth and death rates are regulated under actual conditions. Populations could not endure for a long period of time without this regulation.

Populations are able to last for extended periods of time because of this dependence of the birth and death rates on population size and because of the interaction between organisms occurring in ecological systems.

The most important of such relations are presented above in the form of equations (5.4)–(5.9). These relations characterize the decrease in productivity of excessively large populations due to food shortages, the increase in the death rate of animals of these populations as a result of an increase in the number of parasites and predators, etc. Inverse laws affect the dynamics of small populations.

As a result of the influence of ecological factors, the success of living organisms in the struggle for existence is governed to a significant degree by their ability to adapt to conditions that sustain the stability of ecological systems.

An animal species that possesses a number of advantages in comparison with a species that has adapted to a given ecological system can upset the equilibrium of this system, destroy it, and as a result, be in danger of dying out.

For example, an increase in the size of predators that pursue herbivores and an increase in the aggressiveness of these predators can lead to a destruction of their food objects, and thus to the extinction of the predators themselves. If herbivores develop effective methods of defense against predators, their population may grow and as a result the plant cover of their habitat may be destroyed.

There is some basis for supposing that the criterion that determines natural selection does not necessarily imply an increase in the population of a given species if this increase reduces the stability of the ecological system.

As we noted above, all ecological systems are subject to regular fluctuations of their plant and animal populations. These fluctuations can occur when environmental factors vary or when these conditions are constant (self-oscillatory variation of population). As a result, unstable ecological systems will have a reduced life span and their breakdown will place the populations that compose them in danger of dying out. Highly stable ecological systems have a greater chance of maintaining their populations over a long period of time.

Natural selection thus tends to preserve organisms whose evolution increases the stability of ecological systems. In addition, it supports the existence of more stable systems and eliminates the less stable ones.

Numerical simulation of energy relations in ecological systems leads us to conclude that many systems attain a high stability at early stages of their existence, after which the impetus for further development became highly restricted. An unusually large number of these systems include representatives of ancient forms of organisms in the open sea where environmental conditions have always been more constant.

The rapid course of the evolutionary process which can induce great fluctuations in living organisms is less likely within stable ecological systems in more or less constant environmental conditions. Variations of the organisms of these systems should correlate somewhat, which significantly decreases the rate of evolution for each organism separately.

Nevertheless, even the most stable ecological systems have evolved both as a result of fluctuations in the inorganic medium (for example, due to fluctuations in the gaseous composition of the earth's atmosphere) and as a consequence of purely biological factors. The rate of evolution of such systems is apparently minimal.

The evolutionary process in stable ecological systems is microevolutionary, i.e. it is primarily restricted to the readaptation of low taxonomic groups of living organisms.

This process can also lead to a fluctuation of higher taxonomic groups over sufficiently long periods of time; we are in agreement here with the previously expressed point of view that no sharp boundary exists between micro- and macro-evolution (Mayr, 1963). The rate at which new taxonomic groups spring up, however, is usually comparatively low.

Fluctuations in the environment that disturb the stability of ecological systems have played a substantial role in accelerating the evolutionary process. If these disturbances are sufficiently great, they can bring about the extinction of many groups of living organisms whose ecological niches then remain empty, sometimes for a long time. Less significant disturbances in the stability of ecological systems have led to the extinction of other organisms whose ecological niches were immediately occupied to some extent by new organisms. In these cases, the rate of evolution has increased more rapidly with greater fluctuations of the environmental conditions.

Let us consider an example that illustrates the influence of a fluctuation of external conditions on the succession of fauna.

South America became separated from the other continents shortly

5 CLIMATE AND THE EVOLUTION OF LIVING ORGANISMS 141

after the start of the Tertiary period, so that its fauna developed in isolation for a long period of time. Members of two orders of herbivores, the Litoptera and the Notoungulata, occupied a significant position in this fauna. South America was joined to North America at the end of the Pliocene, and as a result many mammals that significantly differed from the animals in the ecological systems of South America were able to reach the continent. This led to the rapid extinction of many species of fauna of South America, including the overwhelming majority of species of these orders. These species were replaced by tapirs, horses, deer, mastodons, and other animals, which formed stable ecological systems with the predators that had arrived with them from North America (puma, jaguar, bears, and others).

Over nearly the entire Tertiary period the herbivores of South America comprised ecological systems that included predators from the order of Marsupialia, which were apparently less active than the placental predators of North America. It was only natural that the predators of South America became extinct at the same time as the animals they had pursued.

The extinction of many species of widespread animals in South America at the end of the Pliocene is generally explained in terms of the displacement of these animals by more advanced animals from North America (Colbert, 1958). This explanation of the extinction of these animals appears oversimplified from our point of view. A more comprehensive explanation can be seen in the breakdown of the ecological system of South America once its isolation was disturbed.

Using energy balance equations for ecological systems, it is in principle possible to quantitatively calculate the rate of extinction of species and to determine what were the factors that resulted in the extinction of the herbivores of South America, whether it was their pursuit by new species of predators, trophic competition with herbivores from North America, infection by parasites brought by the new groups of animals, or a combination of all these factors.

Let us discuss the role that variations in the size of living organisms play in the course of their evolution. We noted above that the energy flux directed from autotrophic plants to different heterotrophic organisms is the basis for the energy system that includes all living organisms of the biosphere. The energy flux gradually narrows as it travels to new organisms due to great energy losses in each link of the trophic chains. It is completely exhausted when the organic matter of living creatures that complete the food chain is mineralized.

The structure of the energy flux is notoriously nonuniform. A number of links of the trophic chains characterize energy of living matter concentrated in large organisms, while in other links it is distributed in many small living bodies; energy may be transmitted in both directions (for example, herbivorous whales feed on plankton and whale biomass is utilized by different microorganisms).

A highly important evolutionary factor is that of the size of living organisms. This determines their energy level and the amount of energy required to sustain their vital activity. It has been noted for some time that an increase in size can be seen in progressive lines of living organisms (Cope's Law).

The role of size in the ecology of the group plant cover is governed by the preference of large plants for the greatest amount of solar energy in comparison with small plants, which are often extensively shaded. Large plants with long root systems that are part of the noncompact plant cover of arid regions can often obtain water from deeper layers of the soil, while great succulents collect a greater amount of water in their tissues in comparison with the surface from which evaporation occurs. Large plants usually live longer than small plants.

An increase in the body size of animals increases their resistance to changes in thermal conditions, decreases the threat posed by encounters with other animals, and in most cases substantially prolongs their lifetime. In addition, in the case of both plants and animals, an increase in size means an increase in the energy required for their activity, which places a limit on the maximally possible size of living organisms.

It should be noted that the trophic energy obtained by many plants and animals from the environment is proportional (to a first approximation) to their surface area, whereas the energy consumed in activity is proportional to their mass. Since these energy fluxes are equal under average conditions, it is clear that the rate at which energy arrives from the environment per unit surface of an organism will increase in proportion to the size of the organism.

A variation in the energy influx required by living organisms not only limits the maximal size of organisms, but also establishes the optimal size of each organism for the ecological niche it occupies, so that the difficulties in maintaining the energy balance is compensated by the advantages created by an increase in the size of the organism.

We should note that a group of organisms with identical biomass per unit area, but including organisms of different sizes, is characterized by differing sensitivities of each organism to a change in environmental conditions.

Let us consider this question for two animal populations provided with the same food energy base, but significantly differing in size and lifetime. The productivity of the population of smaller animals will be

5 CLIMATE AND THE EVOLUTION OF LIVING ORGANISMS

somewhat greater than the productivity of the population of the larger animals under such conditions, due, in particular, to the higher rate of metabolism of the first group of animals. Calculation using population dynamics equations demonstrate that an increase in the death rate corresponding to the same loss in biomass for small and large animals leads to much more rapid extinction of larger animals than of smaller animals (cf. Chapter 7).

Thus larger animals with a low biomass-reproduction coefficient that possess a number of advantages over small animals under stable environmental conditions are in greater danger when their death rate increases due to changing environmental conditions.

For this reason many large species of organisms in our example vanished, these species being the most vulnerable components of ecological systems whose stability has been disturbed. This probably explains why ecological niches for larger organisms are more often occupied by new, more advanced forms of organisms, whereas many small organisms maintain their ecological niches even when environmental conditions significantly change.

5.2 Climatic Factors in Evolution

Climatic Changes. A change in the composition of the atmosphere has been an important factor in the evolution of living organisms. The enrichment of the atmosphere by biogenic oxygen that occurred in the Proterozoic led to an increase in the rate of metabolic processes and made possible the development of more complex forms of living beings. A subsequent change in the content of carbon dioxide in the air decisively affected the evolution of plants.

As we noted in the preceding chapter, the decreased concentration of atmospheric carbon dioxide that prevailed at the end of the Phanerozoic must have led, other conditions being equal, to a decrease in the mass of autotrophic plants, and consequently, to a decrease in the mass of all living organisms. Under these conditions the evolution of photosynthesizing plants would have accelerated the rate of assimilation of carbon dioxide per unit weight of the plants. Such a change in autotrophic plants can retard the process by which the mass of living organisms decreases as the concentration of atmospheric carbon dioxide decreases.

We should, however, note that this is difficult to prove using data from observations on photosynthesis in contemporary plants from groups of more and less ancient origin. Available data apparently do not indicate a significant difference in the rate of photosynthesis between these groups. However, since these observations refer to contemporary plant forms, they cannot characterize the laws of photosynthesis in the distant past.

Fluctuations in thermal conditions exerted a significant influence on the evolutionary process in the Cenozoic and particularly in the Pleistocene. These fluctuations, as indicated above, led to a temperature drop in all regions of the globe and to a particularly noticeable decrease in the middle and high latitudes. The cooling process accelerated toward the end of the Tertiary and resulted in a sharp temperature drop during the ice ages of the Pleistocene.

The influence of climatic changes on the evolution of terrestrial vertebrates was treated by Mathew (1915). By analyzing paleontological data and data on the contemporary geographic distribution of animals, Mathew was led to conclude that climatic changes exert a substantial influence on the evolution and dispersion of animals.

Mathew assumed that new animal forms appeared in the most recent geological periods principally in the holarctic region, from which they spread to all the continents. He concluded that the widespread theory in zoogeography regarding the previous existence of different land "bridges" between the continents was unnecessary and that the process by which animals have spread can be explained by retaining the contemporary shape of the continents and assuming that they may have been uplifted and sunk within limits corresponding to the depth of the continental shelf zone.

Mathew accepted Chamberlain's hypothesis regarding the mechanism of climatic changes, which states that the increase and decrease in the level of the continents that has occurred from time to time has led to changes in climatic conditions, and that cooling began in high latitudes as the continents rose, and subsequently spread to low latitudes.

Such climatic changes, in the opinion of Mathew, accelerated the evolution of animals and facilitated their spread from the holarctic region into the tropic zone and into the continents of the southern hemisphere.

It should be noted that Mathew's work is now chiefly of historical interest.

Mathew was one of the first to oppose the previously popular theory in biogeography of "continental bridges" which, in the opinion of investigators of the geographical spread of plants and animals, repeatedly arose between different continents and then disappeared. While Mathew's views on this question have been subsequently accepted, his claim that continental shifts did not occur in the past is not accepted by most geologists at the present time. Chamberlain's theory (which Ma-

5 CLIMATE AND THE EVOLUTION OF LIVING ORGANISMS 145

thew accepted) regarding the laws of climatic changes, is clearly outdated. Finally, the theory that the holarctic region was the chief center for the genesis of species in the most recent geological periods is also oversimplified in the light of currently available data.

Works by Axelrod and Baily (Axelrod, 1967; Axelrod and Baily, 1968) are some of the most recent studies of the influence of climatic changes on the evolution of living organisms. Axelrod and Baily in the latter work considered why the dinosaurs became extinct toward the end of the Cretaceous.

In their opinion, continental uplifting occurred in this period, which led to the drying up of many shallow inland seas. As a result, the air temperature in middle and high latitudes significantly decreased and the temperature changes intensified over the course of the year.

Axelrod and Baily surmised that the increase in the range of environmental temperature fluctuations to which the dinosaurs were unable to adapt due to their previous evolutionary development, was the chief cause of their extinction.

In his first work, Axelrod proposed a similar explanation for the well-known extinction of many large mammals during the Pleistocene. Axelrod assumed that the influence of cold winters in high and middle latitudes on the survival of animal calves was particularly substantial and led to the extinction of a number of mammalian species in the ice ages.

Without enumerating other studies of the influence of climatic changes on the evolution of living organisms, let us note that different authors often disagree as to the mechanism of this influence. For example, there exist numerous hypotheses concerning the causes for the extinction of the dinosaurs. Some authors have proposed that an increase in cosmic radiation occurred simultaneously with the extinction of the dinosaurs, or that the air temperature rose beyond the limits that the dinosaurs could endure, which stimulated the activity of mammals who destroyed dinosaur eggs. To a greater or lesser extent, none of these hypotheses are truly justified, and the problem remains unsolved. A number of suggestions regarding a possible method for dealing with this problem are presented below.

Just as many difficulties arise in explaining why many large animals became extinct in the Pleistocene.

It is well known that particularly significant changes in the fauna occurred toward the end of the Pleistocene with the end of the Würm glaciation. At this time the composition of the animal world outside the tropic regions substantially changed. Many previously widespread large herbivores died out in Europe: these included the mammoth, woolly rhinoceros, European steppe bison, and gigantic deer as well as a number of large predators, including the cave lion and cave bear. Substantial fluctuations in the spread of a number of animals that have been preserved also occurred; for example the large flocks of reindeer that had inhabited Western and Central Europe disappeared.

Still more significant fluctuations in the animal world occurred at this time in America, where about 30 species of large animals became extinct.

Other hypotheses besides the decisive influences of climatic changes have been given as to the possible causes of the extinction of large mammals toward the end of the Quaternary.

Different authors even in the nineteenth century expressed the opinion that primitive man may have exterminated the mammoth and a number of other species of animals. This point of view has been maintained in the contemporary literature by Pidoplichko (1963), who connects it, however, to the debatable conception of antiglacialism.

Much data confirming the significant role of man's activity in the extinction of animals at the end of the Pleistocene can be found in studies by Martin (1958, 1966, 1967), who considers that primitive man decisively contributed to the disappearance of many animals not only in the middle, but also in the tropic latitudes.

In one of our works (Budyko, 1967) a numerical model for the biomass balance of animals pursued by Upper Paleolithic hunters was used to explain why many large mammals became extinct toward the end of the Pleistocene.

The results of these calculations (which are set forth in Chapter 7) demonstrate that hunting for large animals could have made these animals extinct within a period of time not exceeding the duration of the Upper Paleolithic.

A similar calculation has since been carried out by Martin (1973) to explain why many large mammals became extinct toward the end of the Pleistocene on the North and South American continents. Using simplified balance equations to describe the biomass of pursued animals, he came to the conclusion that the migration of a comparatively small number of hunting tribes (not more than a thousand men) from Northeast Asia into North America in the Upper Paleolithic could have led to the rapid extermination of all large animals in the region where these tribes originally settled, after which the hunters would begin to migrate to new, still unused territories. In the course of this migration, the hunting tribes pursuing the herds of large herbivores that still remained would move along a wide front from north to south, increasing their numbers. Martin found that Upper Paleolithic hunters could have reached the southern extremity of South America within a comparatively short period of time (about a thousand years), destroying tens of animal species in their wake.

Much attention must be paid to the influence of climatic changes on ecological systems from the point of view that the rate of the evolutionary process depends on the stability of ecological systems.

It is well known that plants and animals are distributed zonally particular geographic zones comprising ecological systems that are similar in many respects.

A close relation between geographic zones and climatic conditions had been established far back by V. V. Dokuchaev. This relation was subsequently studied by Grigor'ev and the author, who compiled a table of geographic zonality (Grigor'ev and Budyko, 1956).

Let us present one version of this table (Table 15). The columns of the table correspond to different gradations in moisture conditions, which are characterized by the aridity index (i.e. the ratio of the average annual radiation balance of the surface of the earth to the heat necessary for evaporating the annual sum of precipitation), while the rows correspond to gradations in the annual values of the radiation balance of the earth's surface. The values of the radiation balance used in constructing this table are for moist surface conditions (Budyko, 1971).

Similar types of plant covers correspond to each column in the table. Forest vegetation predominates at all latitudes under excess moisture conditions, but not under extremely excessive moisture conditions (aridity index below 2/5). In this case forest vegetation is replaced by tundra at higher latitudes and by swamps at low latitudes. (Since an aridity index below 2/5 is observed over more or less extended areas only a comparatively high latitudes, the swamp regions of low latitudes cannot be considered as independent geographic zones.)

Each gradation in moisture is characterized not only by the type, but also by the productivity of the plant cover. In these works, Grigor'ev and the author proposed that the productivity of the natural plant cover grows as the moisture conditions approach the optimal state (for a given radiation balance) and that productivity grows as the radiation balance increases under fixed moisture conditions.

A particular sequence of changes in soil type with substantially similar properties corresponds to each column of Table 15. In particular, the following sequence of soil types is characteristic for excess moisture conditions: tundra, ash-gray and brown forest soil, subtropical red and yellow soil, tropic ash-gray red soil and laterite. The sequence for moderately insufficient moisture is black and dark brown soil, black and

Table 15
Table of Geographic Zonality

Radiation balance of the earth's surface (kcal/cm². year)	Aridity index							
	Less than 0 (extreme excess of moisture)	From 0 to 1					From 1 to 2	From 2 to 3	Greater than 3
		Excess moisture				Optimal moisture	Moderately insufficient moisture	Insufficient moisture	Extreme insufficient moisture
		0–1/5	1/5–2/5	2/5–3/5	3/5–4/5	4/5–1			
Less than 0 (high latitudes)	I Perpetual snow	—	—	—	—	—	—	—	—
0–50 (subarctic and middle latitudes)	—	IIa Artic desert	IIb Tundra	IIc Northern and middle taiga	Id Southern Taiga and mixed forests	IIe Deciduous forest and forest steppe	III Steppe	IV Semi-arid land of moderate band	V Moderate band desert
50–75 (subtropic latitudes)	—	—	VIa Subtropic forest swamps predominate	VIb Subtropic rain forest			VIIa Hardleafed subtropic forests and thickets VIIb Subtropic steppe	VIII Subtropic semidesert	IX Subtropic desert
Greater than 75 (tropic latitudes)	—	—	Xa Equatorial forest swamps predominate	Xb Highly waterlogged equatorial forest	Xc Moderately waterlogged equatorial forest	Xd Light tropic forest and forest savannas	XI Dry savanna	XII Desert-like savanna (tropic semi-desert)	XIII Tropic desert

5 CLIMATE AND THE EVOLUTION OF LIVING ORGANISMS

brown weakly leached subtropical soil, reddish brown subtropical soil, and so on.

A special place in the table is occupied by the region of perpetual snow. This region is characterized by negative or near-zero radiation balance, negative aridity index, and practical absence of plant and soil cover, and also by a subzone of arctic desert with low annual radiation balance and high moisture.

The geographic zonality regularities represented in the table can be explained using the following considerations.

Since the earth is spherical, its surface can be divided into several latitude zones that differ with respect to the radiant energy influx to the earth's surface. Within each zone (other than the perpetual snow zone) various moisture conditions can be observed, from excess to highly insufficient moisture [i.e. different zonally-averaged values of evaporation minus precipitation, *Editor*]. Zones with similar moisture conditions in different latitude bands can be seen to have similar features (which combine into distinct features due to the change in the radiant energy influx). If we successively compare regions in two latitude bands with increasing moisture (or increasing aridity), these similar features will periodically repeat.

A comparison of the annual values of the radiation balance to the mean air temperatures over the growing period shows that these variables are closely related (Budyko, 1971).

Let us consider how geographic zonality conditions vary with climatic changes, for example, with those that occurred in the course of Quaternary glacial development.

Though glaciers developed as a result of particular changes in radiation conditions, as is evident from the data of Chapter 4, simple calculations demonstrate that the radiation balance of the earth's surface outside the ice zone hardly varied as the ice cover shifted to low latitudes. Since the air temperature in middle latitudes noticeably decreased during the ice ages, it is evident that the present-day relation between the radiation balance and the mean temperature of the growing period was disturbed during these epochs.

Many geographic zones may be thought of as having gradually spread to lower latitudes as air temperature decreased. This zone spreading could also have occurred as a result of a variation in the amount of precipitation, which increased in some regions and decreased in other regions as glaciers developed.

Climatic changes in the Quarternary period were slow enough to allow not only animals, but also most plants to shift their geographic ranges into regions with favorable climatic conditions; this is particularly evident from the noticeable change in the plant cover in high latitudes that occurred as a result of the relatively very brief warming of the 1920s and 1930s (Grigor'ev, 1956).

We should, however, bear in mind that the change in geographic zones that accompanied the development of the glaciers was not limited to their spatial shifting. A decrease in air temperature and more or less constant radiation created new types of climatic conditions and, consequently, led to a change of the plant cover and of other components of the landscape in these geographic zones.

Many examples are available in the contemporary epoch of different landscapes in the same geographic zones for different combinations of thermal and radiation conditions.

These differences exist, for example, in regions where the air temperature in the warm part of the year is significantly lower than its mean latitudinal values. In particular, a relatively low summer air temperature in Sakhalin combines with a high radiant energy influx. As a result, the plant cover on Sakhalin is characterized by a number of distinctive features, such as gigantic forms of many plants. These forms are assumed to have appeared as a result of the combination of a relatively high assimilation productivity with high radiation influx and a comparatively low expenditure of organic matter in breathing due to the low temperatures.

This example demonstrates that the shifting of geographic zones as a result of climatic changes has led to fluctuations in many natural processes that form intrazonal ecological systems. In particular, new types of ecological systems apparently arose during the last glaciation over wide middle latitude areas of the northern hemisphere that were not covered by ice.

Limited fluctuations in ecological systems due to the migration and partial rearrangement of the structure of geographic zones may have been accompanied by sharp fluctuations in these systems, when climatic changes would have led to the total disappearance of certain isolated regions of these zones. This may also have occurred on islands, where climatic changes may have resulted in the replacement of forest by tundra or tundra by forest, on coastlines, in regions adjacent to extended mountain areas, etc. Intrazonal ecological systems may have been disturbed under these conditions, and different types of plants and animals whose areas were limited to the corresponding parts of the geographic zone became extinct.

It is, however, likely that the influence of comparatively slow climatic changes on ecological systems led chiefly to their gradual rearrangement, thus accelerating the evolution of living organisms.

This conclusion has been confirmed by Kurten (1965), who established that the rate of evolution of mammals in the Pleistocene significantly exceeded their rate of evolution in the Tertiary.

Short-Term Climatic Changes. The environmental temperature range within which different plants and animals can carry on their vital activity is an important factor in limiting the climatic zone of life.

All plants exist within definite temperature ranges. At low temperatures plants are harmed by ice crystals forming in their tissues. These crystals break down cell walls both directly and by forcing water into tissues as they freeze.

The size of the temperature drop that plants can tolerate varies within a wide limit for different plants. It also depends on the rate at which temperature falls (plants tolerate slow cooling more easily than rapid cooling), and on the degree of hardiness of the given plant, i.e. on the cooling it has previously experienced.

Temperature increases above certain limits also harm plants, for example by destroying chlorophyll and blighting leaves, which lead to the death of the plant if the temperature increases further. The breakdown of different protein compounds in plant tissues apparently plays a substantial role in the mechanism through which overheating harms plants.

Resistance to the effects of overheating differs for various plants; succulent desert plants endure the highest temperatures without being harmed.

Many studies have established that the rate of photosynthesis and the productivity of autotrophic plants substantially depend on temperature, productivity usually approaching zero at high and low temperatures.

Thus, thermal factors exert a profound and decisive influence on the vital activity of plants. The geographic distribution of plants, their seasonal changes, and the species composition and productivity of the plant cover thus depend on the thermal conditions.

The influence of temperature on the vital activity of animals is no less substantial than on plants. All animals can be divided into two groups depending on their thermal conditions, the cold-blooded animals, whose body temperature varies within wide limits, and the warm-blooded animals, whose body temperature is maintained at an approximately constant level. The intermediate group of heterothermal animals, which can regulate the temperature of their body to a limited extent, is sometimes added to these two groups.

The body temperature of cold-blooded animals usually does not coincide with the temperature of the environment because of metabolic processes resulting in the release of heat. However, the heat production of these animals is, as a rule, insignificant, so that these differences are usually comparatively low. Exceptions to this rule related to moments of high activity by the animal are possible. For example, the body temperature of many insects sharply increases during flight.

There exist other factors whose effect can facilitate a change in the temperature of cold-blooded animals relative to the environment. This temperature can decrease due to hat loss from evaporation from the body surface of the animal and due to long-wave radiation into the surrounding space. The absorption of solar radiation can lead to an increase in body temperature.

Fluctuations in body temperature for large cold-blooded animals (for example, some reptiles) can lag noticeably behind fluctuations in the temperature of the environment because of the thermal inertia of the animal's body.

The warm-blooded animals include two higher classes (birds and mammals) that have developed an effective mechanism in the course of evolution for maintaining a constant body temperature independent of external conditions.

This mechanism is controlled by a heat regulation center, the hypothalamus, found in one section of the brain. As the external factors of the thermal conditions vary, this center correspondingly stimulates the physical and chemical heat regulation systems, which allows the animal to maintain a constant body temperature. The wide variability of metabolism, which increases sharply as temperature decreases, is the principal means of temperature regulation. Fluctuations in evaporation from the body surface, the state of the hair coat, etc. are other means of temperature regulation in many animals.

Body temperature differs for different warm-blooded animals. It is usually between 40°C and 42°C for birds and often between 36°C and 39°C for mammals. The constancy of the body temperature of warm-blooded animals is relative, and it often fluctuates somewhat during the day with variations in muscular tension, as a function of feeding conditions and other factors. An animal's body temperature can fluctuate noticeably with external environmental conditions that are anomalous for the particular warm-blooded animal.

Both cold and warm-blooded animals carry out their vital activity within a given range of body temperature, limited by what is called the "lethal" temperature. This temperature may differ significantly for the same species of animal as a function of the animal's acclimatization, i.e. on its prior exposure to increased or decreased temperatures.

The lethal temperature of cold-blooded animals is highly varied. Some animals die when their body temperature fluctuates compara-

5 CLIMATE AND THE EVOLUTION OF LIVING ORGANISMS 153

tively slightly, while others tolerate temperature fluctuations of up to tens of degrees without harm.

The difference between the upper and lower lethal temperatures for warm-blooded animals often is 15–25°C. The lower lethal temperature for a human being with normal body temperature of about 37°C is between 24–25°C, and the upper lethal temperature is 43–44°C.

The biological functions of both cold-blooded and warm-blooded animals vary significantly as their body temperature fluctuates. The vital activity of cold-blooded animals is highly susceptible to the influence of temperature, such activity being possible only within a fixed and sometimes quite narrow temperature range. Thus the behavior of cold-blooded animals is often determined by the need of maintaining an optimal body temperature. For example, many insects, reptiles, and other animals will move to areas illuminated by the sun and assume a position providing the greatest radiant heating of their body in the case of insufficient heat. These animals will leave the surface of the earth when it becomes extraordinarily heated by the sun, either burrowing or climbing up the branches of plants where the air temperature is below the temperature of the earth's surface. We should particularly note the important effects of thermal conditions on the breeding process of cold-blooded animals, which is governed in many ways by their ecology in the corresponding period.

Though the ecology of warm-blooded animals depends to a lesser extent on the temperature conditions, under unfavorable conditions these animals will also use the microclimatological features of the landscape in order to bring their body temperature to an optimal level.

These actions are particularly necessary in the case of small animals whose bodies possess insignificant temperature inertia. In addition, they cool very easily at low environmental temperatures due to the high ratio of body surface, through which heat is lost, to body mass, which determines heat production as a result of metabolism.

The substantial influence of temperature factors on the vital activity of animals makes the thermal conditions of the environment an important factor of the geographic distribution of animals and their ecology. Obviously, the geographic distribution of different plants and animals depends substantially not only on the mean climatic conditions, but also on deviations of weather conditions from their norm.

Data from meteorological observations demonstrate that the air temperature at the surface of the earth may differ by tens of degrees from their mean monthly value for a given region at different moments of time. Great temperature anomalies (particularly negative anomalies) often lead to systematic destruction of plants and animals.

The influence of great weather anomalies on the number of animals was noted as far back as 1859 by Charles Darwin, who described a sharp decrease in the number of birds following a severe winter in England. In his book, "The Origin of Species," Darwin wrote (pp. 68–69), "The action of climate seems at first sight to be quite independent of the struggle for existence; but in so far as climate chiefly acts in reducing food, it brings on the most severe struggle between the individuals, whether of the same or of distinct species, which subsist on the same kind of food. Even when climate, for example extreme cold, acts directly, it will be the least vigorous individuals, or those which have got least food through the advancing winter, which will suffer most." It is clear from this remark of Darwin that systematic destruction of animals during highly anomalous weather can occur long before the meteorological elements reach their lethal values.

It has been established in climatological studies that great anomalies in meteorological elements often encompass areas extending over thousands of square kilometers. In many cases the areas of the anomalies exceed the areas occupied by different plants and animals. Thus sufficiently great anomalies in weather conditions can lead to the extinction of several species.

Since the size of a probable anomaly in air temperature, like a number of other meteorological elements, increases with increasing period of time, it is obvious that fluctuations in climatic conditions could have influenced living organisms in a very substantial way over the long epochs of the earth's geological history.

The influence of short-term climatic changes on the evolution of living organisms has been considered in one of our previous works (Budyko, 1971), where it was noted that the evolutionary process may have depended not only on slow climatic changes, but also on short-term, sharp fluctuations in climatic conditions.

This possibility is a consequence of the enormous time scales of nature's development in the geological past, scales beyond those that can be directly observed by man.

In the preceding chapter we considered short-term climatic changes caused by volcanic activity. These changes very probably exerted a definite influence on the progression of fauna.

It is well known from paleontological studies that the process by which higher taxonomic groups of animals (orders, suborders, and families) became extinct is characterized by the fact that biologically distinct groups of animals that occupied entirely different ecological niches often all became extinct within short (from the standpoint of geological history) time intervals. Statistical concepts imply that this phenomenon cannot have been a result of the struggle for existence under invariant external conditions.

5 CLIMATE AND THE EVOLUTION OF LIVING ORGANISMS   155

As an example let us consider the well-known simultaneous extinction of the large groups of animals towards the end of the Cretaceous period. At this time, five of the 10 existing orders of reptiles vanished, including the dry-land dinosaurs (two orders), the pterosaur, and also ichthyosaur and Plesiosaurus. These five orders included 35 families containing a great number of species (Colbert, 1965).

Let us assume that the mean duration of existence of the reptile orders that became extinct in the Cretaceous was roughly 100 million years, and let us also assume that the time at which an order disappeared can be determined to within 5 million years.

The probability that five out of 10 orders will randomly die out within 5 million years is roughly one in 10,000 as calculated using Poisson's formula. Such an extremely small probability indicates that the extinction of these groups of reptiles was not a random coincidence, but was due to definite fluctuations in nature.

In view of the paleontological data, it is easy to verify that the higher taxonomic subdivisions of the animal world in most cases did not disappear one by one, but in comparatively large groups. In particular, most of the orders and suborders of reptiles and amphibians that became extinct vanished within three critical epochs of geological history, toward the end of the Permian, the Triassic, and Cretaceous. Calculation shows that such a coincidence would have been highly unlikely for constant or slowly varying external conditions, even if these critical epochs lasted several million years.

It is as well impossible to explain the disappearance of many large groups of animals by the struggle for existence under invariant external conditions because of a number of other qualitative concepts besides the above calculations.

It has been repeatedly noted in the paleontological literature that mammals occupied the ecological niches of the reptiles of the Mesozoic, not competing with these reptiles, but only following their extinction, when the corresponding niches were free.

We may recall from a number of other examples of this type the interesting case of the extinction of the phytosaurs at the end of the Triassic. These rptiles suddenly disappeared, freeing the corresponding ecological niche, which was appropriated after some time by the crocodiles who had sprung from the same initial forms as the phytosaurs and who were extremely similar to them. It would be quite difficult to explain such examples without taking into account fluctuations in natural conditions.

Thus to understand the causes for great fluctuations in fauna in the geological past, it is necessary to take into account the effective factors associated with sharp fluctuations in environmental conditions as well as the influence of purely biological factors.

The enormous effect that may have been exerted by short-term temperature drops on different forms of organic life cannot be doubted. As we noted above, all living organisms without exception are characterized by a definite temperature zone within which they exist, which in many cases is very narrow. A sharp decrease in temperature lasting for several years or tens of years will lead to the destruction of many species of plants and animals, primarily those habituated to climatic conditions near the lower boundary of their temperature zone. While a decrease in temperature over the entire planet would allow various species to continue living in the warmest regions, many plants and animals would re-establish their previous geographic ranges once a temperature reduction ceased. But if cooling of the warmest regions was sufficiently great, a given species or group of species would completely vanish.

We should note that the actual methods by which sharp temperature drops affect the vital activity of different organisms vary widely.

For every organism we may establish the size of a temperature drop that may (a) lead directly to its destruction; (b) decrease its activity down to the limits at which it perishes in the course of its struggle for existence; (c) decrease the resistance of the organism to infectious illnesses, as a result of which it becomes their inevitable victim; (d) disturb the breeding process.

The temperature drops corresponding to these conditions may differ, though even the least of them will be sufficient to cause the extinction of some species of animals.

The extinction of the reptiles and amphibians toward the end of the Triassic, which was outstanding in its scale, and the still greater extinction of reptiles toward the end of the Cretaceous, are the most interesting simultaneous disappearances of large groups of animals from the geological past.

Before passing to a discussion of these two milestones in geological history, we must above all emphasize that both occurred during a period of intensive orogenesis. This has been repeatedly noted in the geological literature (Strakhov, 1937).

The probability that a large number of volcanoes will simultaneously erupt is extremely high during a period of active orogenesis. We should bear in mind that the significant general intensification of volcanism was accompanied not only by an increase in the probability of a particularly great temperature drop, but also by an increase in the probability of a series of such drops, which could have been grouped over hundreds of thousands or millions of years.

It is well known that one important feature of the reptiles and amphibians was their lack of temperature regulation, which made these animals particularly susceptible to thermal conditions.

The climate of the Mesozoic was characterized by a weakly expressed thermal zonality, a higher temperature in the low latitudes than is the case in the contemporary epoch, and a much higher temperature outside the tropic regions, which created favorable conditions for the existence of cold-blooded animals lacking temperature regulation over the entire earth.

A different situation held towards the end of the Paleozoic, when the air temperature over extensive areas of the earth's surface decreased from time to time, which apparently created difficulties for cold-blooded animals. Peculiar features developed by some reptiles of the Permian (dimetrodon, edaphosaur) illustrate this in an interesting way. These reptiles carried a large crown on their spine, which, in the opinion of some paleontologists (Romer, 1945), was used to heat their body in the case of low temperatures. By positioning this crown perpendicular to solar rays, these reptiles were able to receive an additional amount of heat sufficient for further physical activity.

The absence of such an adaptation in Mesozoic reptiles is one indication of a comparatively warm climate at all latitudes during this epoch. Furthermore, sudden and significant temperature drops must have exerted a pronounced effect on cold-blooded animals of that time.

Reptiles that still exist are the vestiges of a rich and varied fauna from the Mesozoic era which have survived after the extinction of the overwhelming majority of the large (and many small) forms of this fauna. There are grounds for assuming that contemporary reptiles have a greater resistance to temperature drops than those which have died out, so that their ecology may provide some idea of the upper limit of the resistance of reptiles to cold.

The comparatively small number of reptiles whose habitat is poleward of the subtropical zone are active only in the warm time of the year and protect themselves in the cold periods by different methods. Many reptiles have typical daily rhythms of activity which usually cease at the coldest hours of the day.

Works by Bogert (1959) and his colleagues, who have studied the ecology of reptiles, established that reptile activity is enormously dependent on the necessity of maintaining a fixed body temperature. When heat is insufficient, reptiles will attempt to get warm using solar rays or the heat from the earth's surface. When cooled, some lizards change their color in order to absorb more solar radiation.

We may reliably conclude, using data from laboratory and field observations, that a decrease in the mean temperatures in the habitat of many reptiles of about 10° will lead to their destruction. The ecology of one animal belonging to the most ancient of the still existing orders of reptiles (Rhynchocephalia), is thus of particular interest. The only species occurring in this order is the tuatara, which now inhabits several islands in New Zealand. The existence of this animal under the cold climatic conditions of New Zealand is sometimes considered evidence for the comparatively high resistance of ancient reptiles to low temperatures (Bellairs and Garrington, 1966).

However, such a conclusion is based on an incorrect understanding of the climate of New Zealand. It is a pronounced maritime climate distinguished chiefly by the absence of any significant chills during the year, an extremely favorable factor for cold-blooded animals.

Recent studies have established that the embryo in the tuatara's egg matures very slowly and that this process continues for 15 months; the embryo ceases development during the coldest time of the year. Such a long period greatly increases the risk of destruction of the tuatara's eggs by other animals, which demonstrates how difficult it is for ancient types of reptiles to survive at insufficiently high temperatures. Evidently, even a rather small temperature drop will make it impossible for the tuatara's eggs to develop and will lead to the extinction of this species.

It may be that a sharp temperature drop could have affected the cold-blooded animals of the Mesozoic. However, it is now difficult to establish the precise criteria governing the sensitivity of different, now fossilized reptiles to cold and causing some forms to become extinct while preserving others. We can only assume that a temperature drop will have been particularly destructive to large animals and also to oceanic species, which could not have sheltered themselves in warmer spots at moments of the greatest chills, as is done by contemporary reptiles.

This hypothesis reasonably explains the nearly total extinction of the large land reptiles as well as both orders of marine reptiles toward the end of the Upper Cretaceous.

Though the comparatively small forms of land reptiles had a greater chance of surviving because of their ability to find shelter, only in limited, very warm regions could several groups of reptiles have survived following sharp chills.

It is entirely obvious that sharp chills were far less dangerous to animals with temperature regulation, i.e. mammals and birds. These animals are known to have successfully endured the transition from the Cretaceous to the Tertiary and have since occupied all the ecological niches of the reptiles that had meanwhile become extinct.

5 CLIMATE AND THE EVOLUTION OF LIVING ORGANISMS 159

The degree to which the extinction of the reptiles at critical points in their history was in fact simultaneous is of some interest. Though the analysis of available paleontological data encounters great difficulties due to the unreliability of dating different deposits, it appears that the different species of reptiles became extinct toward the end of the Upper Cretaceous within comparatively close but not coinciding periods of time. This conclusion can be made to agree with theories of the influence of sharp, short-term cooling on the extinction of reptiles if we bear in mind the probability of a series of short-term temperature drops during periods of increased volcanism. The first drops would destroy animals least adapted to cold, while succeeding drops would lead to the extinction of other groups weakened by the earlier coolings.

Let us now discuss the degree to which this theory allows us to overcome the difficulties mentioned in some studies in explaining the extinction of Mesozoic reptiles. We will consider here an idea of Colbert, who set forth the modern viewpoint on this question in th concluding section of his book, "The Age of Reptiles" (Colbert, 1965).

In discussing the possibility for the influence of catastrophic fluctuations in external conditions on the extinction of reptiles, Colbert (pp. 203-204) wrote, "Did some great event take place that wiped out these reptiles? If so, why should a great catastrophe be so selective—why should it have caused the extinction of five orders of reptiles and allowed the continuation of five other orders? Why should all families and genera and species belonging to these five orders have disappeared completely? Moreover, if it was a physical catastrophe of some sort, why is it not recorded in the rocks? The continuity of sediments to be seen in certain localities as Cretaceous beds are succeeded by Tertiary beds indicates an orderly world in those distant days, like the orderly world we know. And indeed this is an expression of the principle of uniformitarianism, so important to the science of geology. From all of the records of the rocks, the world around, there is ample reason to think that earth processes and life processes in the past were as they are today; that events of past ages must be explained by the forces which we see acting in Nature at the present time. There is no place for world-wide catastrophes in the world of the past or of the present if th principle of uniformitarianism has any validity."

Colbert further noted that the Cretaceous ended in the "Laramide Revolution," a term referring to the formation of a number of contemporary mountain systems. He assumed that this change in natural conditions was quite slow and that reptiles could have adapted to it in the course of biological evolution. Colbert proposed that the difference in the climate of the Upper and Lower Cretaceous periods was extremely small and could not have caused the extinction of the reptiles. In discussing the possible importance of a temperature change in the transition from the Cretaceous to the Tertiary, Colbert was chiefly drawn to the hypothesis that temperature sharply increased at this time. He indicated that it was difficult to make this hypothesis agree with the fact that some groups of reptiles became extinct while others were preserved. Colbert particularly doubted that a temperature change could have influenced the extinction of marine animals, since the ocean possesses a high thermal inertia.

Colbert also rejected such hypotheses as competition with mammals able to destroy the reptile eggs, the exposure of the reptiles to radiation, etc. He deduced that the extinction of the reptiles was caused by a combination of complex phenomena whose true nature will probably never be explained.

From our point of view, such a pessimistic conlcusion can be considered justified only on the basis of paleontological data and without resorting to methods and data from related disciplines. Recourse to data of paleoclimatology allows us to overcome the difficulties indicated by Colbert and to answer the questions he posed.

In particular, this theory implies that sharp temperature drops whose destructive influence on cold-blooded animals is doubtless must have occurred during a period of time comparable to the duration of the Mesozoic era. These drops were most likely in the period of active volcanism, when the extinction of the reptiles also occurred. Calculations show that the upper layers of the ocean were significantly chilled during these coolings, and this led to the destruction of marine species of reptiles.

Because of their short-term nature such coolings could not leave direct traces in geological deposits, though their effect on the organic world would have been catastrophic in the full sense of the word.

As we noted above, the qualitative laws governing the extinction process, according to which only land and chiefly small forms of reptiles were preserved, reasonably agree with the likely consequences of short-term cooling. The extinction of all members of some orders and the preservation of some members of other orders could reflect to a certain degree the different resistance of members of different orders to cooling. The direct relation between the evolutionary origin of reptiles in different orders and the influence of the temperature conditions on them is quite natural.

Thus the largest fluctuations in the stock of living organisms apparently occurred as environmental conditions changed sharply. The example of such changes given above far from exhausts their possible mechanisms, which require further study.

Fluctuations in environmental conditions, and particularly great changes in these conditions, could have significantly increased the rate of evolution.

This conclusion agrees with Simpson's viewpoint (Simpson, 1944), who noted that the extinction of ancient groups of animals often coincided with periods of mountain formation. He wrote, ". . . among reptiles and mammals the proportion of orders that seem to have arisen during times of pronounced emergence and orogeny is greater than can reasonably be ascribed to chance. This is probably true of birds, also . . ." (p. 113). Simpson also tried to draw attention to the decreased rate of evolution in tropical latitudes, where external conditions were always more stable.

In his work, Simpson isolated a specific type of the evolutionary process he called "quantum evolution."

In quantum evolution, the higher taxonomic units, families, orders and classes are replaced, and the rate of variation of living organisms in this case attains the highest value. We must assume that the mechanism of quantum evolution is of great importance when ecological systems become unstable, that is when their organisms, in the course of experiencing rapid changes, partly die out and partly become components of new ecological systems.

Thus climatic changes have exerted a definite influence on the evolutionary process, this influence apparently being particularly significant in epochs when short-term sharp climatic changes intensified.

5.3 Climatic Factors in Human Evolution

Human Evolution. The hypothesis that man is descended from ape-like ancestors was expressed even in the eighteenth century. I. Kant advocated this hypothesis and it was supported by J. B. Lamarck at the beginning of the nineteenth century.

The works by C. Darwin and A. Wallace, published at the end of the 1850s and containing a theory of the origin of the species by means of natural selection of the most adapted organisms, were of great importance for the explanation of the origin of man. Though the problem of human evolution was not considered in these works (Darwin limited himself to the remark that his theory will shed new light upon human evolution), it immediately became obvious that the Darwin-Wallace theory could be used to explain human evolution.

As early as 1860, a famous dispute broke out between T. Huxley and S. Wilberforce in which Huxley defended the hypothesis that man originated from ape-like animals as a result of the process of natural selection. Works by T. Huxley, E. Haeckel, and K. Vogt and other authors that developed this point of view were published in the 1860s.

A somewhat different position was taken by Wallace. Though he expressed viewpoints similar to those of Huxley in his first work dealing with human origin (Wallace, 1864), he came to the conclusion after several years that natural selection could not have been the only basis for the origin of man (Wallace, 1869). In the opinion of Wallace, the intellectual abilities of man would only slightly surpass the abilities of the higher apes as a result of natural selection. He noted that the development of the human brain significantly exceeded the requirements of human existence at earlier stages of human society. In order to account for this discrepancy, Wallace allowed for the intervention of a "higher intelligent will" in the process in which man originated.

It is worth noting that, though Wallace was interested in the spiritualism much in style during the second half of the nineteenth century, the hypothesis that nonmaterialistic factors influenced the origin of man was the only such assumption taken in all his many biological studies (George, 1964).

Darwin's opinion regarding human evolution was expressed in his book, "The Descent of Man and Selection in Relation to Sex," published in 1871. This book contained voluminous data demonstrating that man arose from ape-like ancestors similar in several features to a number of contemporary higher apes. Darwin suggested that the likely spot at which man appeared was Africa, where there now exist two apes most similar to man, the chimpanzee and the gorilla (the same opinion was expressed by Wallace).

In his book Darwin only fleetingly referred to Wallace's concepts regarding the difficulty of explaining the development of the human brain and indicated that he disagreed with these theories.

This lack of agreement was apparently based on a comparison Darwin carried out between the intellectual activity of man and animals. In fact, he found much in common between them. As a result of this comparison, Darwin was led to conclude that there was no qualitative difference between animal and human intelligence. Such a conclusion can hardly be considered justified from the standpoint of contemporary concepts.

It is worth noting the conclusion with which Darwin ended his book. "We must, however, acknowledge, as it seems to me, that man, with all his noble qualities . . . with his god-like intellect which has penetrated into the movements and constitution of the solar system—with all these exalted powers—Man still bears in his bodily frame the indelible stamp of his lowly origin" (Darwin, 1971).

It is clear from this sentence that Darwin recognized to some degree the fundamental differences between man and the most advanced animal forms.

Though Wallace outlived Darwin by several decades and witnessed many new advances in the study of the origin of man, he nevertheless preserved his opinions on this question to the end of his life.

The difference of opinion between Wallace and Darwin on such an important question is of some significance in view of the scientific authority Wallace held and the fact that he highly valued Darwin's studies and only in rare cases expressed himself against his conclusions. Nevertheless it was quickly forgotten, probably as a result of Wallace's clearly unacceptable hypothesis that nonmaterialistic factors affected the evolutionary process.

Only very little paleontological data illuminating the origin of man were available in the middle of the nineteenth century. Many outstanding discoveries have been made in this field over the last hundred years, the most important of them being the discovery of the remains of Australopithecus in southern and eastern Africa as well as the remains of Pithecanthropus and beings related to him (Homo erectus) in Java, China, and Africa.

Data from paleontological studies have confirmed the Darwin–Wallace viewpoint that man arose in Africa, though many features of Australopithecus, who was similar to the ancient ancestors of man, differ from contemporary higher apes. Australopithecus was to a significant degree a carnivore, and hunted different animals using the simplest tools. Thus the use of animal meat, which is characteristic of modern man, has a very ancient origin.

Australopithecus existed toward the end of the Pliocene and throughout the Quaternary, i.e. within an interval of time from several million years to several hundreds of thousands of years before our era. Members of the species Homo erectus, who occupied an intermediate position between Australopithecus and man in many respects, lived for several hundred thousand years in the Pleistocene.

Modern man (Homo sapiens sapiens) appeared about 40,000 years ago, though it is assumed that modern man is descended from Neanderthal man (Homo sapiens neanderthalesis), who existed for a longer period of time and vanished approximately when modern man appeared. Since later forms of Neanderthal man differ from modern man to a greater extent than do earlier forms, the origin of modern man from later Neanderthal man has often been subject to doubt.

The greater part of time during which modern man has existed is encompassed by the first culture he created, the culture of the Upper Paleolithic. During this period hunting for large animals was the economic foundation of life in human society. It was concluded in the studies whose results are set forth in Chapter 7 that the culture of the Upper Paleolithic ended in an ecological crisis encompassing great areas, due to the destruction by Upper Paleolithic hunters of the animals they pursued.

As we noted above, this crisis was a highly unusual phenomenon in the history of the biosphere, since ecological systems as a rule regulate the population of organisms that are part of particular food chains, making rapid extinction of species of plants and animals impossible. Disturbance of this regulation indicates that the emergence of modern man constituted a definite boundary, and in passing this boundary man was no longer governed by the biological laws that determine the size of animal populations.

The extermination of many large animals apparently led to a reduction in the population of several extensive regions up to the time when hunting and gathering of useful plants began to be replaced by cattle breeding and agriculture, the latter occurring in the Neolithic epoch, about 10,000 years ago. Thus the period during which man has produced the food he uses amounts to only about one-fourth of the total duration of his existence.

Industrial civilization developed within an even briefer period of time — a fraction of a per cent of all human history — while the scientific and technical revolution takes up only some hundredths of a per cent of human history.

Though available paleontological and archaeological data illuminate many stages in human evolution, the mechanism of this evolution requires further study.

One difficulty in explaining this mechanism was understood by Darwin, who wrote in his conclusion regarding the origin of man from ape-like ancestors, that "the high standard of our intellectual powers and moral disposition is the greatest difficulty which presents itself, after we have been driven to this conclusion on the origin of man" (Darwin, 1871, p. 609). To overcome this difficulty it is necessary to clarify the question posed by Wallace over one hundred years ago.

It is well known that natural selection generally preserves changes in organisms that are useful for their vital activity and which are employed in the course of their existence. However, it appears that the vast abilities of the human brain have not been utilized to any great extent throughout most of the entire history of modern man.

There are grounds for believing that the central nervous system of modern man has scarcely changed since the Upper Paleolithic. The paintings of animals in caves of Western Europe, which are often considered the greatest achievements of animal painting over all of human history, is one proof of this fact.

The impression that an enormous gap exists between the abilities of modern man in the pursuit of intellectual activity and the use of these abilities at previous stages of cultural development arose from observations on transmitting the achievements of contemporary civilization to members of the most backward tribes, some at the cultural level of the Stone Age. Since they easily enter modern industrial civilization if they receive the proper education from childhood, it follows that their intellectual abilities were not used to any great extent under the previously existing conditions.

It is highly likely that the greatest achievements of modern civilization are still much below the level modern man could reach were he to completely use his intellectual abilities. This possibility further widens the gap between man's potential abilities and their use throughout his history.

An explanation of this gap within the framework of the ordinary theory of evolution by means of natural selection of the most adapted organisms encounters a number of difficulties. Though many animals master certain forms of behavior as a result of domestication that are not natural to them in the wild state, there exist no analogies for this gap in the animal world. The hypothesis that so complex and perfect an organ as the brain of modern man could have appeared as a result of preadaptation, i.e. to some degree by accident, is clearly unlikely.

Thus an explanation of the origin of man requires that we take into account the effect of particular factors that led to man's qualitative difference from all animals. The nature of these factors warrants special attention.

The Role of Climatic Changes. We may suppose, following the viewpoint of a number of researchers (Gerasimov, 1970), that climatic changes that occurred in the Pliocene and Pleistocene exerted a particular influence on human evolution. As we noted in the preceding chapter, in this time cooling developed in middle and high latitudes, which led to significant climatic changes over all the continents. It is highly likely that these climatic fluctuations accelerated human evolution as well as the evolution of many other organisms.

Any quantitative estimate of the rate of human evolution meets with special difficulties. According to some data (Kurten, 1965), the mean duration of existence of European species of mammals in the Pleistocene amounted to about one million years. This value is comparable with the duration of human-like species in the Pleistocene. Thus the rate at which species that were ancestors of man were replaced does not appear to be very high.

However, we should assume that such an estimate of the rate of evolution is quite tentative.

The accepted system of classification of living organisms does not reflect their ecological features to any great extent. It is well known that this system is based mainly on morphological and phylogenetic principles. However, there are many examples of ecologically similar organisms that are highly distinct in origin. In addition, closely related organisms often possess entirely different interrelations with the environment.

The assertion that the higher apes are morphologically more similar to man than to the lower apes was presented as one argument favoring this theory during the first studies of the origin of man based on the idea of natural selection.

It seems that the difference between modern man and Neanderthal man was greater from the ecological point of view than that between Neanderthal man and many mammals that do not belong to the order of primates. The basis for this difference is man's vast influence on the environment, which manifested itself even during the Upper Paleolithic (when man exterminated tens of species of large animals) and has reached an enormous scale in recent times. Though Neanderthal man lasted roughly ten times as long as has modern man, he did not disturb the balanced ecology of the biosphere and in this respect was far more similar to other mammals than to modern man.

There is now a tendency to reduce the rank of taxonomic groups that divide man from the primates related to him. This tendency, which is entirely justified from the standpoint of the principles by means of which the classification of living organisms is constructed, amplifies to an even greater extent the lack of correspondence between this classification and the differences between the ecological properties of man and his ancestors.

We must conclude, bearing in mind the scale of this difference, that changes in human ecology occurred much more rapidly than the mean rate of change of the ecology of different animals.

This process may have been due to the development of the human central nervous system, whose elements are not taken into account to any great extent within the classification of man and beings related to him. Available data demonstrate that the size of the brain of man's ancestors nearly tripled over two million years (Kurten, 1971). At the same time, the brain structure became more complicated.

One cause for such a rapid development of the central nervous system may possibly be the critical circumstances of man's predecessors in the tropics during the Pliocene and Pleistocene, when noticeable climatic changes occurred. These changes were a consequence of cooling at high latitudes, which led to an increase in the mean meridional gradient of the air temperature. As a result, the atmospheric circulation system changed and the high pressure belt shifted to lower latitude. Since little atmospheric precipitation falls within this belt, the moisture conditions varied over large tropical regions, which led to the replacement of tropical forests by savannas and semideserts in a number of regions.

There exist grounds for supposing that prior to this climatic change man's ancestors inhabited forests and hardly differed from other higher primates except for a more developed erect posture. It is highly probable that, like all primates, they were basically herbivores.

The direct ancestors of Australopithecus (who may be provisionally called pre-Australopithecus) were morphologically similar to Australopithecus and were comparatively small (the body weight of ancient Australopithecus was at most 30 kg). The rate of travel of erect pre-Australopithecus was much less than for most four-legged animals of that size, and they also lacked such organs of attack and defense as large canine teeth and claws.

Pre-Australopithecus was almost defenseless in open country conditions, where the possibility of concealment in trees was often absent, even when attacked by small members of the many predators of the African savannas. In addition, the disappearance of many edible plants he fed on in the tropical forests significantly limited his supply of plant food.

The low birth rate of pre-Australopithecus, which is characteristic for the higher primates, and the great length of time for the development of young, who were a particularly easy catch for predators pursuing pre-Australopithecus, made it considerably difficult to maintain pre-Australopithecus populations.

The chief factors which could have insured the survival of pre-Australopithecus included their comparatively highly developed brain and their erect posture, which completely freed their hands.

Pre-Australopithecus was exposed to conditions of exceptionally intense natural selection, which led to a high rate of evolution of this small population. Such an evolutionary process, called quantum evolution by Simpson, does not usually leave traces in the paleontological records due to the small number of animals involved (Simpson, 1944).

Paleontological studies have shown that Australopithecus mastered the technique of applying the simplest tools for defense and attack. Using these tools, Australopithecus succeeded in attacking such large apes as the baboon and, possibly, different hoofed animals. As a result, they were able to catch a significant amount of animal food.

A number of discoveries of Australopithecus remains demonstrates that their populations were not too small and that, consequently, they were able to overcome the crisis situation discussed above.

It is worth noting that groups of comparatively small Australopithecus were in a difficult situation when attacked by large predators even if they used such tools as stones, sticks, and bones of hoofed animals. It was no less difficult for the slowly moving Australopithecus to catch any rapidly moving animal. Their success in the struggle for existence could have come about only as a result of the significant advantage provided by their intellectual abilities in comparison with all the animals who attacked them or which could have been their prey. Natural selection in the course of the evolution of Australopithecus was therefore enormously favorable to the development of a brain, which made possible a gradually increasing success against predators and in hunting for different animals.

An increase in the intellectual abilities of Australopithecus made him more capable of using different objects as tools and made possible a transition to the production of tools. It was of enormous importance that Australopithecus found it necessary to engage in collective actions for self-defense and when attacking other animals. Elements of social organization subsequently arose out of these collective actions, which led to the development of moral concepts to define the behavior of the individual in a collective. The importance of the labor process and the development of social relations for human evolution was clarified in a well-known work by F. Engels. (F. Engels, The role of labor in the process by which apes became man, The Dialectics of Nature (in Russian), Moscow, 1950, pp. 132–144). Engels also drew attention to the important role of the process in which man's ancestors switched to animal food, which significantly accelerated man's physical development.

Paleontological data indicate that some lines of Australopithecus attained enormous size in the course of their evolution. This significantly decreased the pressure of natural selection on the evolution of the nervous system and led to a loss of the habits of tool use. These lines came to an evolutionary dead end and their members died out.

Lines of man's ancestors characterized by progressive development of the central nervous system existed alongside such regressive lines. This development was probably not uniform. It accelerated when the

5 CLIMATE AND THE EVOLUTION OF LIVING ORGANISMS 169

pressure of natural selection on the populations of man's ancestors intensified. It is highly likely that such epochs coincided with fluctuations in natural circumstances and with the development of Quaternary glaciations.

Since climatic fluctuations encompassed enormous areas in the course of the glaciers' advance and recession, these fluctuations could have significantly varied the conditions under which man's ancestors lived, which set new problems before them and accelerated their evolution.

The last great stage in human development apparently took place during the Würm, when Homo sapiens sapiens appeared.

These conceptions describe the role of climatic changes in the origin of man, though they are not sufficient for deciding why the human brain developed beyond the limits determined by the needs of primitive society.

From our point of view, this problem can be explained if we take into account the discrepancy between the crude means available to man's ancestors and those tasks it was necessary for them to solve.

Until recently hunting for large animals and protection against predators involved significant difficulties and risk even for civilized peoples with firearms. These difficulties were incomparably great for man's ancestors, who used primitive tools for hunting, and particularly for distant ancestors, who were significantly smaller and weaker than modern man.

Nevertheless, man's ancestors inevitably encountered large predators. Dangerous circumstances arose also when man's ancestors gathered animal food, that is, when only large animals could supply a sufficient amount of meat for more or less numerous tribes.

Undoubtedly, man's ancestors as a rule were the victors in such encounters. This is confirmed, in particular, by discoveries of the remains of such enormous animals as elephants, mammoths, and cave bears at campsites of ancient man.

Man's ancestors, when equipped with primitive tools, were able to successfully fight large animals only because of highly developed special habits far exceeding the habits that highly equipped modern man finds necessary for successful hunting.

A paradoxical situation arises. The less perfected were the hunting tools available to man's ancestors, the greater the degree to which they compensated for the disadvantages of these tools by a high level of intellectual activity, which permitted them to get the better of encounters with large animals. Therefore, enormous demands on the intellectual development of man's predecessors were apparent in the distant past, which led to the crystallizing of progressive changes in the development of the brain by means of natural selection.

On the basis of this theory, it appears that human evolution accelerated in those regions where hunting for large animals was the chief source of food. Such conditions prevailed, in particular, in the vast forestless regions adjacent to ice covers in the ice ages. It is possible that modern man emerged in the middle latitude region.

Modern man could have appeared under ecological conditions that placed very stringent demands on his intellectual activity both for successfully combating large and dangerous animals, and, possibly, in order to come out the victor in competition with related primates.

Following the emergence of the subspecies Homo sapiens sapiens, man's material culture progressed more rapidly in regions where the natural conditions did not provide him with plant food, but required hunting for large animals.

Members of this subspecies, which populated all the continents, appeared in a number of cases in zones where food gathering was aided by the availability of easily accessible edible plants (humid tropics and subtropics) or of animals that could not strongly resist hunters. The latter case probably occurred when Australia was colonized by modern man. Marsupials were an easy prey for the experienced hunters of the Upper Paleolithic. Nearly the same thing may have happened when America was first colonized, the various animals there never having come into into conflict with man and lacking protective instincts against hunters (Martin, 1973). Under these conditions, cultural development was retarded and previously attained advances may have sometimes been entirely lost.

These theories imply that enormous reserves of intellectual activity were freed as material culture, which insured man's struggle for existence, developed. Such reserves, when stimulated by complex environmental conditions, could be used to further increase the level of material culture. Thus culture spontaneously developed, and at some stage made possible the emergence of the first elements of civilization. The development of social relations in human society was of fundamental importance in this process.

This progress was highly nonuniform in space and time. Whenever environmental conditions gave rise to great stresses on human society, the use of the potential intellectual abilities of man reached a high level. This level significantly decreased under the comparatively favorable conditions which were far more frequent.

These thoughts can explain the difficulties involved in the explanation of the origin of man and pointed out by Darwin and Wallace.

6 Man's Influence on Climate

6.1 Changes in Local Climate

Influence on Plant Cover. The influence of man on climate began several thousand years ago as a result of the development of agriculture. Forest vegetation was destroyed in many regions in order to cultivate the land, which led to an increase in the wind speed at the earth's surface, a change in the temperature and moisture conditions in the lower layer of the atmosphere, and also to a change in the soil moisture, evaporation, and river discharge. Destruction of forests in comparatively arid regions was often accompanied by an increase in dust storms and destruction of the soil cover, which significantly changed the natural conditions in these areas.

In addition, the destruction of forests even over extended areas apparently exerted a limited influence on large-scale meteorological processes.

A decrease in the irregularity of the earth's surface and a fluctuation in evaporation on areas free of forest somewhat varied the precipitation regime, though this variation was comparatively low if the forests were replaced by other types of plant cover (Drozdov and Grigor'eva, 1963).

Complete destruction of the plant cover in an area, which repeatedly occurred in the past as a result of man's agricultural activity, exerted a more substantial influence on precipitation. These cases were particularly frequent following the felling of forests in mountain regions with a weakly developed soil cover. Under such conditions erosion rapidly destroyed soil not protected by the forest, as a result of which the further existence of a mature plant cover became impossible. Similar circumstances arose in some regions of dry steppes, where the natural plant cover, destroyed as a result of unlimited grazing of agricultural animals, was not restored; these regions became deserts.

Since areas of the earth's surface lacking a plant cover are intensively heated by solar radiation, the relative humidity over these areas falls, which increases the level of condensation and decreases the amount of precipitation. It is possible that such a mechanism is of importance for explaining cases in which the natural vegetation in arid regions was not renewed after being destroyed by man.

Planting of forests is also accompanied by changes in weather conditions which, however, are basically limited to the lowest layers of the atmosphere. Protective forest strips exert the greatest influence of all the different types of forest plantings on climate and are widely used as means of reclamation.

Forest strips are usually in the form of plantations from several meters to several tens of meters wide, which border square or rectangular fields measuring from several hundred meters to between one and two kilometers.

Forest strips are often grown in drought-stricken regions where they are used to maintain more favorable moisture conditions in farm lands.

The wind-breaking effect of forest strips, resulting from the decrease of the mean wind speed in fields between the strips and from the decrease in the rate of turbulent exchange near the earth's surface, is the principal way in which they influence the meteorological regime of the lowest air layer.

The attenuation of turbulent exchange in fields between the strips is explained by the fact that vortices moving near the earth's surface are broken up to a significant extent after passing through the forest strip. As a consequence, the vortices will lack large eddies, which substantially decreases the intensity of eddy motions in this flow (Yudin, 1950).

It should be noted that such an effect is observed only if the vortex has passed through the forest strip comparatively unimpeded. A thick, impenetrable forest strip acts in a completely different manner. A comparatively small calm zone is created behind such a strip, beyond which the wind rapidly picks up speed and approaches the conditions of the open steppe. In this case decreases in the size of the eddies are not observed in the lowest layer.

These phenomena are explained by the fact that an air stream rises somewhat as it approaches an impenetrable forest strip, arches over it, and then immediately descends and returns roughly to its original state.

A decrease in the intensity of the turbulence in the lower air layer over fields between the strips is of great practical importance. Data from recent studies show that turbulence directly affects the development of two meteorological phenomena, the blowing of snow from fields and the formation of dust storms. A decrease in turbulent exchange near the earth's surface safeguards against or weakens dust storms and helps to maintain snow on farm lands between the strips.

6 MAN'S INFLUENCE ON CLIMATE

A decrease in turbulent exchange is also of substantial importance for preserving moisture in the soil during the warm part of the year.

The amount of possible evaporation (evaporative power), as well as a number of other meteorological factors, depends on the rate of turbulent exchange in the lowest air layer. Previous calculations (Budyko, 1971) have demonstrated that evaporative power decreases by roughly 10% when the mean values of the exchange coefficient in the lowest air layer decrease by 20%.

Besides decreasing evaporative power, protective strips help to increase snow reserves on fields and increase the amount of precipitation. The influence of all these factors leads to a significant increase in soil moisture in fields protected by strips.

As noted above, numerous observations have shown that the run-off of snow waters significantly decreases in fields with protective strips. This decrease is chiefly explained by the change in the distribution of the snow cover on protected fields from that on unprotected fields. Abatement of the wind speed and of turbulent exchange in the lowest air layer in fields within the strips create conditions for a comparatively uniform distribution of the snow cover, whereas a significant part of the snow in open fields is blown away in gullies and other topographic depressions, and after thawing simply drains away. Moreover the increased permeability coefficient of the soil under strips impedes to a greater extent the snow melt in comparison with open fields, which also decreases the spring run-off of snow water.

A fluctuation in the amount of precipitation due to a variation of vertical velocities in the atmosphere and of evaporation may also exert some influence on the water balance of soil protected by strips.

We may use heat and water balance equations, taking into account the values of the integral diffusion coefficients, run-off, and precipitation corresponding to the conditions of protective strips in order to estimate the total influence of these hydrometeorological conditions on the water balance of the soil (Budyko, 1956).

Calculations have demonstrated that soil moisture significantly increases and evaporation somewhat increases over protected fields. The increase in soil moisture depends on the season and the conditions of turbulent exchange, run-off, and precipitation.

Since a system of afforestation strips, besides impeding run-off of snow melt, decreases turbulent exchange in summer and simultaneously increases precipitation, soil moisture increases not only at the beginning of the planting season, but also during its second half.

But if (as often happens) the influence of forest strips is chiefly manifested in an increase of snow reserves and a decrease of spring run-off ("winter effect"), soil moisture will increase only in spring and at the beginning of summer.

The amount of productive moisture attained in the soil under these conditions can be several tens of per cent greater than in unprotected fields, other conditions being equal.

These·conclusions are satisfactorily confirmed by data from observations in regions with mature penetrable forest strips.

A significant increase in the amount of free moisture in the soil and an increase in total evaporation can substantially improve the yield of agricultural crops under average climatic conditions. Yields may improve as a result of an increase in the productivity of plant transpiration (which is aided also by a decrease in exchange and wind speed in the lowest air layer). In addition, the ratio of the amount of water consumed in transpiration to total evaporation noticeably increases with increasing soil moisture, which also enhances the productivity of agricultural crops as total evaporation grows.

Thus the use of forest strips can significantly vary the water balance in the soil and noticeably increase productivity, which, as is well known, is confirmed by data from numerous experimental studies.

Forest strips, in addition to influencing the water balance in the soil, also play an important role in abating dust storms which drastically damage the soil cover in arid regions in some years. This was clearly manifested in the 1968-69 winter when powerful dust storms passed through the southern part of the European territory of the USSR. Inspection showed that the winter crops on protected fields suffered significantly less damage than on other fields.

Forest strips are also used in regions of sufficient humidity, where they increase the mean temperature of the earth's surface during the warm part of the year by attenuating turbulent mixing. Under such conditions they exert a favorable influence on the development of thermophilic crops and accelerate the maturation of many agricultural plants (Gol'tsberg, 1962). Without dwelling on other methods of changing the meteorological regime, we should note that the effect of man on the plant cover changes climatic conditions in the lowest air layer within certain limits.

The Modification of the Water Regime. One way by which man affects climate involves his use of artificial irrigation. Irrigation has been utilized in arid regions for thousands of years, beginning with the most ancient civilizations in the Nile valley and the Tigris-Euphrates interfluvial area.

Irrigation sharply varies the microclimate of irrigated fields. A significant increase in heat loss by evaporation leads to a decrease in temperature and an increase in the relative humidity of the lower air layer. This variation in the meteorological regime, however, rapidly weakens beyond the irrigated fields. Irrigation leads only to fluctuations in the local climate and does not really affect large-scale meteorological processes.

Let us discuss in detail the physical mechanisms by which irrigation affects the meteorological regime (Budyko, 1956). The radiation balance substantially increases when irrigation is used in arid steppes, semideserts, and deserts; this increase may reach several tens of per cent of the initial balance.

This change in the radiation balance can be explained, on the one hand, by the increase in the amount of absorbed short-wave radiation as a consequence of a decrease in albedo, since the albedo of moist soil covered by more or less abundant vegetation is significantly less than that of a desert or semidesert. On the other hand, a decrease in the temperature of the underlying surface and an increase in the humidity of the lower air layer with irrigation decreases the effective radiation, which also increases the radiation balance.

Irrigation under arid climatic conditions leads to a sharp increase in heat losses in evaporation, whose magnitude is governed principally by the irrigation rates. An increase in the loss of heat in evaporation as a rule exceeds the increase in the radiation balance under ordinary irrigation rates. As a consequence, turbulent heat transfer sharply decreases and becomes negative, corresponding to the direction of the mean turbulent heat flux from the atmosphere to the underlying surface in the case of sufficiently high irrigation rates. This is manifested in the appearance of daily temperature inversions.

Thus irrigation under arid climatic conditions significantly decreases both the turbulent heat flux (which can even change its direction), and the heat flux transmitted by long-wave radiation. This may lead to sharp fluctuations in air mass transformations over a given region when irrigating sufficiently large areas.

Fluctuations in the components of the heat balance in the case of irrigation can be estimated from the observations at Pakhta-Aral (central Asia). Results of these observations depicted in Fig. 30 allow us to compare the heat balance components of an irrigated oasis and its surrounding semidesert for summer conditions.

The data in Fig. 30 demonstrate the radiation balance R noticeably increased in the oasis in comparison with the semidesert regions and that a large heat loss in evaporation from irrigated fields occurred (evaporation over the investigated period of time was practically zero in the semidesert region). Accordingly, the turbulent heat flux P in the

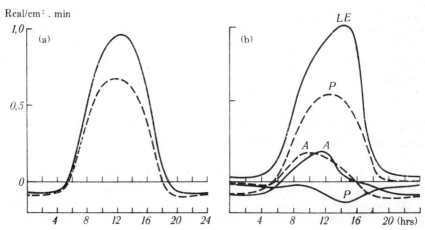

Fig. 30. Daily variations of the heat balance components in an irrigated oasis (solid line) and semidesert (broken line). (a) Radiation balance; (b) heat loss on evaporation (LE), turbulent heat flux (P), and heat exchange in the soil (A).

desert region can be seen from Fig. 30b to be much greater than the flux in the oasis during the day and has an opposite sign. The flux of heat in the soil for these conditions changes comparatively little with irrigation.

We noted above that irrigation exerts a substantial influence upon the thermal regime. Due to low heat losses in evaporation in deserts and steppes, solar radiation absorbed by the earth's surface is basically consumed in atmospheric heating by means of turbulent heat transfer and long-wave radiation. Under these conditions, extremely high ground surface temperatures are observed. To calculate these temperatures, we may use the heat balance equation in the form

$$R_0 - 4\delta\sigma T^3 (T_w - T) = \rho c_p D (T_w - T) + LE + A, \qquad (6.1)$$

where R_0 is the radiation balance of the earth's surface determined by calculating the effective radiation with respect to air temperature; $4 \delta\sigma T^3 (T_w - T)$ is the difference between the effective radiation determined from the temperature T_w of the earth's surface and the air temperature T; δ is a coefficient that characterizes the difference in the properties of a radiating surface from the properties of a black body, σ is the Stefan-Boltzmann constant, $\rho c_p D (T_w - T)$ is the turbulent heat flux from the earth's surface to the atmosphere, ρ is air density, c_p is the specific heat of the air for constant pressure, D is the integral coefficient of turbulent diffusion, LE is heat consumption by evaporation, L is latent evaporation heat, E is evaporation, and A is the heat influx to the soil.

Equation (6.1) implies that

$$T_w - T = \frac{R_0 - LE - A}{\rho C_p D + 4\delta\sigma T^3}. \tag{6.2}$$

Since LE and A are much less than R_0 under these conditions, $T_w - T$ can reach high values in daytime. Calculations using (6.2) have shown that it may be as high as 10–20°C.

Large values of $T_w - T$ correspond to strong heat fluxes from the earth's surface to the atmosphere. The temperature in the lower air layer increases as a result of heating and the relative humidity falls. A decrease in the relative humidity in turn decreases the amount of precipitation.

The water balance of the soil changes considerably when drought-stricken areas are irrigated. Evaporation sharply increases as a result of irrigation, this increase being equal to the irrigation norm minus the loss of irrigation water due to infiltration. Heat consumption in evaporation correspondingly increases, which leads to a substantial decrease in the temperature of the earth's surface.

Since the air temperature fluctuations observed in irrigation are significantly less than the temperature fluctuations of the earth's surface, we may write a formula for the temperature fluctuations of the earth's surface in the course of irrigation:

$$T_w - T_w' = \frac{(R_0 - R_0') - (LE - LE')}{\rho C_p D + 4\delta\sigma T^3}, \tag{6.3}$$

where the values referring to irrigation are indicated by a prime. In deriving equation (6.3) it was assumed that $|A - A'| \ll |LE' - LE|$.

We indicated above that though the radiation balance of the earth's surface strongly grows in the course of irrigation due to a decrease in albedo, the increase in heat consumption in the course of evaporation significantly exceeds the increase in the radiation balance for ordinary irrigation rates.

As a consequence, the difference $T_w - T_w'$ is quite high for sufficiently high irrigation rates, the temperature of the earth's surface over the irrigated section approaching the air temperature under daytime conditions; it is less than the air temperature in the case of abundant irrigation. We noted above that this leads to large changes in the conditions of air masses transformation when sufficiently great areas are irrigated, as a result of which the temperature and humidity conditions of the lower air layers change over irrigated areas. Dry warm air entering from outside is humidified and cooled as it moves over an irrigated area.

Temperature and humidity fluctuations of the air over irrigated sections can be considered as results of the transformation of the air arriving from outside in different directions. The magnitude of these fluctuations depends on distance from the border of the irrigated oasis. The further is the given section from the border of the oasis, the greater will be the fluctuations. In addition, temperature and humidity fluctuations of the air with irrigation depend on a number of factors, including:

1) Irrigation rates and time intervals between irrigations. The greater the amount of water an irrigated field receives and evaporates over a given interval of time, the greater will be the temperature and humidity differences between irrigated and nonirrigated fields.

2) Wind speeds and coefficient of turbulent exchange. These two closely related factors govern the thickness of the air layer in which temperature and humidity fluctuations occur. The temperature and humidity distribution with altitude within the lower air layer chiefly depend on these factors.

3) Radiation properties of the underlying surface, chiefly its reflectivity (albedo); we have spoken of this factor above.

Any empirical study of the influence of all the factors taken separately is extremely difficult so that it is more suitable to calculate theoretically the temperature and humidity fluctuations and to verify the results by means of data of stationary and field experimental studies (Budyko, et al., 1952).

On the basis of theories that have been developed, we may calculate temperature and humidity fluctuations at different altitudes with a satisfactory precision as a function of water loss in evaporation, fluctuations of albedo, rate of turbulent exchange, and distance from the oasis edge. These calculations can be used to plan irrigation systems over previously nonirrigated areas.

Let us pass to a consideration of empirical data.

Table 16 presents data generalized by S. A. Sapozhnikova (Budyko et al., 1952) of observations carried out at weather stations located in a desert and in irrigated oases. The data are reduced to a single latitude (42°) and altitude over sea level (100 m). Oases whose width did not exceed 3 km were called small, while the other oases were called large. It is evident from the table that fluctuations in temperature T and humidity e were maximal in absolute value during summer months when irrigation water discharge was greatest. At this time the temperature drop at the level of the weather station due to an increase in evaporation and transpiration was 2.5°–3°C, while the absolute humidity at the center of a large oasis was roughly 5 $mbar$ greater than in the desert. At lesser altitudes, the absolute values of the differences grew.

Table 16
Effect of Irrigation on the Meteorological Regime

	Apr.	May	June	July	Aug.	Sept.	Oct.	Nov.
Temperature difference:								
small oasis − desert	0.0°	−0.5°	−1.6°	−2.4°	−2.5°	−1.7°	−1.4°	−0.4
big oasis − desert	−0.6°	−1.1°	−2.2°	−3.1°	−2.8°	−2.3°	−1.7°	−0.8
Humidity difference (mb)								
small oasis − desert	1.1	1.8	3.4	3.6	3.7	2.5	1.2	0.4
big oasis − desert	0.4	1.8	4.2	5.4	5.4	3.6	1.6	0.8

The influence of irrigation on air temperature and humidity is somewhat less in a steppe zone, mainly due to a lower additional water discharge through evaporation.

It should be noted that irrigation of drought-stricken areas, which decreases the temperature over irrigated fields, increases the mean atmospheric temperature. A temperature drop in irrigated areas results from an increase in heat consumption for evaporation, but this growth for the earth as a whole is compensated for by an identical increase in the heat influx due to condensation; this heat is released in the atmosphere over other regions where water vapor created by irrigation is condensed.

In addition, the albedo of the earth's surface significantly decreases when desert and arid steppes are irrigated, which increases the amount of radiation absorbed in the earth-atmosphere system.

Drainage of marshland usually has an effect on the climatic conditions that is opposite to that of irrigation. The soil temperature increases and evaporation decreases due to the decrease in soil moisture, tillage can begin earlier, harvesting in autumn is facilitated, etc. (Budyko, Drozdov, and Yidin, 1966).

Furrow ploughing, which can be used in the north when there is not enough heat, is one method of drainage and thereby of warming water-logged soil. The soil in the plough horizon becomes 1°–1.5° warmer than in level fields as a result of this form of tillage.

A natural peatbog exhibits a number of properties that differ from those of mineral soils. In the moist state peat contains a great quantity of water and therefore possesses a high heat capacity with significant heat conductivity. In contrast, dry peat and the mossy vegetation covering it is characterized by low heat conductivity. The dry surface layer of peat, since it is strongly heated by day and cooled at night, transmits little heat to the soil underlayers. Therefore under such peat layers frost is retained for a long time at the beginning of summer, sometimes disappearing only after rain moistens the soil. In summer, the peaty soil of a bog is relatively cold, while in winter, on the other hand, the bog hardly freezes, and often thaws out under the snow cover in the course of the winter. Evaporation from undrained swamps varies within wide limits. If the peak is thin (as is usually the case in low-level grassy bogs), the bog will be covered by thick grassy vegetation in summer, from which a quantity of water that is near the maximally possible for these energy resources will evaporate (given moisture reserves are sufficient). Grass-land is usually replaced by moss when a thick layer of peat is present (usually high bogs). Moss easily absorbs water, but evaporates it comparatively weakly, particularly if the bog is covered with shrub vegetation. The total evaporation from such bogs depends on the moisture saturation of the bog. According to Romanov (1961), bogs in the northern European territory of the USSR are saturated with moisture and evaporate several tens of percent more than dry valleys. The mean yearly evaporation from bogs in the central part of this territory is not greater than evaporation from dry valleys, while in the most arid regions it may be even less.

The bog climate also varies depending on the manner and purposes of its reclamation. Thus bogs are thoroughly drained in order to extract peat. Due to the low capillary conductivity of peat, the surface of the bog easily dries up, vegetation on it dies, and evaporation substantially decreases when it is thoroughly drained. Daily temperature fluctuations increase sharply over a dry peat surface.

Drainage of marshland is carried out in an entirely different way for the purpose of converting it into agricultural land, the microclimatological conditions varying correspondingly with reclamation. In this case reclamation encompasses only the surface layers. Under these conditions evaporation from vegetation is determined by the energy resources as well as by the state and phase of development of the vegetation. In general, evaporation approaches evaporative power during vegetation periods.

No such increase in daily temperature ranges noted over parched marshland is observed over a moist peat surface. At the same time the soil temperature in drained marshland under thick swards is less than the temperature of arid valley loamy soil by 3–6° and is less than the temperature of dry valley sandy soil by 4–8°.

Large water storage basins on rivers created in the course of constructing hydroelectric power stations constitute large but comparatively shallow reservoirs. Therefore their influence on a climatic change is similar to that of small reservoirs. Their main effect is to decrease the irregularity of the earth's surface and correspondingly increase wind. The wind speed over water reservoirs is several tens of per cent greater than over open level country. This increase is greatest in autumn, when the water is warmer than the air and when intensive turbulent exchange develops over reservoirs, and is comparatively weakly expressed in spring both when the reservoirs are covered by ice and immediately after it has melted when the water is cold. At this time, the wind speed over water basins is about the same as over open level country. Daily fluctuations of the air temperature decrease following the construction of a water basin, the radiation balance increases (as a consequence of a decrease in the albedo of the area), and the mean yearly evaporation increases, while it is distributed differently than over dry land in the course of a year.

The fluctuation in environmental climatic conditions is not great when a water storage basin is built in a location with excess or sufficient moisture. For example, it is difficult to note any systematic temperature fluctuations in the vicinity of the Rybinsk water storage basin at any point along its present shores.

Substantially larger climatic changes arise when reservoirs are located in regions characterized by insufficient humidity (Tsimlyanskoe, Volgogradskoe, Bukhtarminskoe reservoirs, etc.). The temperature along the coastlines of these basins is noticeably less (by 2–3°) in the warm time of the year than in regions far from the water, due to the greater evaporation from a reservoir than from the surrounding dry land (where the rate of evaporation is limited by low soil moisture and from which dry air arrives at the coastal parts of the basin). A daytime air temperature drop promotes the development of rather strong (up to 3–4 m/sec) breezes extending vertically to several hundred meters.

Artificial reservoirs, like irrigation, decrease the earth-atmosphere system albedo and, consequently, increase the amount of absorbed radiation. A water basin thereby increases the mean temperature of the atmosphere. This increase is obviously less than the temperature change resulting from irrigation.

Urban Climate. Climatic conditions in cities are usually more variable than in the surrounding regions; these variations are greater, other conditions being equal, the larger is the area of the city.

Available data indicate that climatic changes in large cities appeared hundreds of years ago. Thus, Landsberg (1956) presented eyewitness evidence regarding the severe air pollution in London in the seventeenth century; such air pollution significantly attenuated solar radiation in the city as compared with agricultural areas.

The chief factors influencing the meteorological regime of a city include:

1) a fluctuation in the albedo of the earth's surface, which is usually less in built-up areas than in a rural area;

2) a fluctuation in mean evaporation from the earth's surface, which is noticeably less within the boundaries of a city (though immediately after rainfall evaporation from roofs and pavements may exceed evaporation in a rural area);

3) heat release created by different types of man's economic activity, whose amount may be comparable to the amount of solar radiation incident upon the area of a city;

4) an increase in the irregularity of the earth's surface within a city as compared with a suburban area;

5) pollution of the atmosphere by different solid, liquid, and gaseous impurities created in the course of economic activity.

One of the chief features of the urban climate is the appearance in a city of a "heat island," which is characterized by greater air temperatures than a rural area. This effect has been investigated in many experimental works, and several numerical models have been proposed for its study.

Results of previous investigations (Inadvertent Climate Modification, 1971; Landsberg, 1970; Berlyand and Kondrat'ev, 1972) have demonstrated that a heat island is usually of a complex structure, each city block of an urban development being a source of heat for surrounding undeveloped sections. The mean air temperature in a large city is often 1–2° greater than the temperature of surrounding regions, though the temperature difference may reach 6–8° at night with a light wind. This difference usually decreases with strong winds.

Interesting observations characterizing the growth of a heat island in the process of urban development were carried out by Landsberg and Maisel (1972). Construction was begun in 1967 in the USA, near Washington, D.C., of the city of Columbia. In 1968, when only a small part of the city had been constructed, the air temperature in its center was about 0.5° greater than the suburban area at nighttime. In 1970, after significant progress was made in its development, the temperature increase at the center of the city reached 4.5°, and over a large part of its area, 2°. The temperature increase was noticeably less in the daytime, when the intensity of air movement was greater.

The relative humidity decreased by several percent over the area of Columbia at the same time as the temperature increased, this decrease apparently being due to both the temperature increase and a decrease in evaporation throughout the city.

Evidently, a heat island appears chiefly as a result of the influence of the first three factors given above, which govern urban climatic conditions. The relative role of each of these factors may strongly vary in different cities and in different seasons.

It should be noted that not only the relative, but also the absolute air humidity usually decreases in heat islands due to the decrease in evaporation in built-up sections.

Additional heating of the air over cities creates local circulation systems that resemble sea breezes, and also amplifies ascending convective motions over cities. In addition, an increase in the surface irregularity leads to a noticeable decrease in wind speed in cities in comparison with suburban areas.

Air pollution from different impurities, which in many cities reaches high levels, has the greatest practical importance of all climatic features. Discharges from industrial plants, heating systems, and traffic are the source of these impurities.

Some of the anthropogenic aerosol in cities is formed from discharges of solid and liquid particles, while the rest of it is due to gases mixed with the atmosphere.

An increase in aerosol concentration over cities sharply decreases the solar radiation arriving at the earth's surface. According to Landsberg (1970), direct solar radiation in large cities often decreases by about 15%, ultraviolet radiation, on the average, by 30% (it may completely vanish during winter months), and the period of solar irradiation, by 5–15%. The thin atmospheric boundary layer, which contains the greatest amount of aerosol particles, plays the chief role in the attentuation of solar radiation in cities. Horizontal visibility is usually sharply diminished within this layer, sometimes to 10–20% of its value in rural regions.

The high concentration of aerosol particles in urban air increases the frequency of fog and, particularly, of smog—a type of stable fog whose droplets contain a significant amount of impurities that pollute the atmosphere. In some cities (Los Angeles is a particularly well-known example) where local conditions reduce atmospheric circulation, smog may remain for many days, greatly harming the health of the population.

Urban fog plays an important role in attenuating solar radiation as well as in decreasing the range of visibility in cities.

A decrease in solar radiation in cities and industrial regions with a very high level of air pollution can lead to a decrease in air temperature in the daytime. This decrease sometimes completely compensates the increase in temperature due to the heat island (Berlyand, et al., 1974).

The increased quantity of condensation nuclei in the air over cities and the intensification of ascending air movements lead to an increase in cloudiness and precipitation. Available data indicate that there exists a week-long cycle in the amount of precipitation in some industrial centers, which can be explained by a decrease in precipitation on weekends, when the industrial plants are not in operation.

Some increase in precipitation and a significant decrease in evaporation lead to an increase in run-off, facilitated in many cities by surface water sewer systems.

Snow removal and its transportation outside the city are often used in cities with great amounts of solid precipitation, which accelerates the increase in the temperature in the city in the spring in comparison with the suburbs.

Urban climate may be significantly improved by rationally situating residences and industrial plants and by creating green belts and reducing air pollution.

There are many examples that illustrate how a change in the heating system (conversion from solid fuel to gas or to electrical energy) sharply decreases the pollution of urban air and fluctuations of related climatic elements. Of no less importance is the removal of industrial plants from the city, the use of an effective method of air purification in smoke stacks and other sources of pollution, the creation of large parks within a city, and landscaping streets and various sectors that have not been developed.

At the contemporary level of development of industry, energetics, and transport, urban construction cannot be carried out without a detailed study of the influence of man's economic activity on the meteorological regime. Calculations of possible changes in local climate are necessary for proper city planning in order to prevent any climatic conditions that could present a hazard to the health of the population.

It is worth noting that a number of features of urban climate have spread over extensive areas reaching hundreds of kilometers as urbanization has progressed in regions with high population density. Under these conditions, many closely situated and densely populated areas act as a source of heat and air pollution, and their effect is cumulative. This situation occurs in a number of regions in the USA, Japan, and Western Europe, where local climatic changes encompass ever-greater areas.

6.2 Changes in the Global Climate

Atmospheric Carbon Dioxide. Data have been obtained over recent years indicating that man's contemporary economic activity influences not only the local climatic conditions of different regions, but also the climate of our planet as a whole. Variations in the amount of carbon dioxide in the atmosphere is one of the factors involved in this influence.

The hypothesis that the mass of carbon dioxide in the atmosphere began to increase as a result of burning large amounts of coal, petroleum, and other types of fuel was expressed as early as the first half of our century (Callender, 1938). However, only in the 1950s, in conjunction with the advent of the International Geophysical Year, were systematic observations begun of atmospheric carbon dioxide at a number of stations, which permitted a quantitative estimate to be made of the growth in its concentration.

These observations demonstrated that significant annual fluctuation of carbon dioxide at the earth's surface (decrease in concentration in summer as a result of the intensification of photosynthesis) is accompanied by a stubborn tendency toward a growth in concentration from year to year. According to observational data at Mauna Loa (Hawaii), the yearly increase in the concentration of carbon dioxide amounted to $0.64 \cdot 10^{-4}\%$ between 1958 and 1968. This magnitude corresponds to approximately 0.2% of the amount of carbon dioxide contained in the atmosphere (Inadvertent Climate Modification, 1971).

Results of observations at Mauna Loa agree with data from similar observations in Alaska, Sweden, and Antarctica (Bolin and Bischof, 1970). Since these observations were carried out at stations situated at great distances from each other, there is no doubt that they indicate a real tendency in the variations of the atmospheric carbon dioxide concentration.

Data of these observations imply that approximately half the carbon dioxide created in the course of human activity is retained in the atmosphere; the other half is apparently absorbed by the ocean, and, to a lesser degree, by living organisms.

The mechanism by which the additional carbon dioxide is absorbed by water reservoirs and living organisms has not been adequately studied. Though the oceans have a potentially great capacity and can absorb an enormous amount of carbon dioxide, the actual role of absorption of carbon dioxide by sea water is significantly reduced due to the slow rate of exchange between the surface and deep layers of the ocean.

The rate of photosynthesis increases as carbon dioxide concentration increases, though the additional amount of organic matter created here is mineralized within a limited time, releasing the carbon dioxide consumed in the process.

The construction of a complete quantitative theory that would take into account the effect of buffer processes in the ocean and biosphere on variations in carbon dioxide concentration in the atmosphere is a task for the future. At the present time we may use empirical schemes for this purpose, one example of which is the simple numerical model proposed by Machta (Inadvertent Climate Modification, 1971).

Calculations using this model imply that the amount of carbon dioxide in the atmosphere has increased by 0.003% over the last century, i.e., by roughly 10% of its contemporary concentration; the chief part of this increase in the content of carbon dioxide occurred during the most recent decades.

Using the results of recent calculations of the effects of CO_2 concentration on the air surface temperature, we find that its increase caused by man's economic activity could have increased the mean global temperature of the earth's surface by 0.3–0.4°.

This value is comparable to the temperature fluctuations that have occurred over the past century. Thus it is highly likely that the burning of different types of fuel exerts a definite influence on contemporary climatic conditions.

Atmospheric Aerosol. We noted above that a large number of particles, noticeably increasing the concentration of atmospheric aerosol, enters the atmosphere as a result of man's economic activity.

Interesting observations on the fluctuation of dust content in the atmosphere were carried out by Davitaya (1965), who used data on the vertical distribution of dust concentration in the snow cover of Caucasian glaciers.

These data indicated that the amount of dust in a unit volume of the upper snow layers significantly exceeded that in the deeper layers. In the opinion of Davitaya, this difference corresponded to the sharp increase in dust concentration in the atmosphere that had occurred over the preceding decades.

McCormick and Ludwig (1967) as well as other authors have presented data indicating a decrease in the transparency of a clear atmosphere over the last several decades, apparently due to the increase in the concentration of atmospheric aerosol.

Available estimates (Inadvertent Climate Modification, 1971) indicate that about 200 million to 400 million tons of anthropogenic aerosol is now ejected into the atmosphere, which comprises 10–20% of the total amount of aerosol particles that enter the atmosphere. Only a small portion of this quantity is ejected into the atmosphere in the form of solid or liquid particles; their chief sources are the gaseous mixtures created by man, such as sulphur dioxide, nitric oxide and others, from which aerosol particles are formed as a result of various chemical reactions.

These observational data indicate that anthropogenic aerosol noticeably increases the total concentration of aerosol particles not only locally in cities and separate industrial regions, but also over large areas, which confers a global nature on the contemporary process of atmospheric pollution.

The influence of stratospheric aerosol on the air temperature conditions at the earth's surface was considered in Chapter 3. Humphreys had already established that this influence is basically determined by a decrease in the flux of short-wave radiation entering the troposphere.

Aerosol particles in the stratosphere play the role of a screen that changes the meteorological solar constant to a greater or lesser degree. Absorption of radiation by these particles can cause local heating of the stratosphere, as occurred in particular, after the eruption of the Mount Agung volcano (Newell, 1971), though this heating scarcely affects the thermal regime at the earth's surface due to the insignificant air density in the stratosphere and weak heat exchange between the stratosphere and the trophosphere. Thus, as has been demonstrated in a number of studies, an increase in aerosol concentration in the stratosphere invariably leads to a temperature drop at the earth's surface.

Aerosol particles in the troposphere exert a more complex influence on thermal regime. These particles attenuate the flux of short-wave radiation arriving at the earth's surface as a result of back-scattering and absorption of radiation by aerosol particles.

While the first process increases the albedo of the earth-atmosphere system, the second may decrease it.

Vertical redistribution of absorbed radiation within the trophosphere, where vertical heat exchange is intensive, influences the mean temperature of the trophosphere and the mean air temperature of the earth's surface to a comparatively small extent. Therefore the influence of tropospheric aerosol on the thermal regime is basically determined by the dependence of the albedo of the earth-atmosphere system on aerosol concentration, an increase in the former leading to a temperature drop at the earth's surface, while its decrease, to a temperature increase. Obviously, a change in the albedo of the earth-atmosphere system due to aerosol will depend on the albedo of the earth's surface. The lower is this albedo, the more likely is it that atmospheric aerosol will increase the system albedo. The probability that the earth-atmosphere system's albedo will decrease because of aerosol will grow for higher surface albedo (snow, ice).

Numerical models have been constructed in several studies to describe the influence of aerosol on the radiation and thermal regimes of the atmosphere (McCormick and Ludwig, 1967; Charlson and Pilat,

1969; Atwater, 1970; Barrett, 1971; Yamamoto and Tanaka, 1971; Rasool and Schneider, 1971; Mitchell, 1971; Ensor, et al., 1972; Newman and Cohen, 1972; Kondrat'ev, et al., 1973). Some of these studies took into account only the influence of aerosol on back scattering of short-wave radiation, a circumstance which leads to an increase in the albedo of the earth-atmosphere system, and, consequently, to a decrease in air temperature at the earth's surface.

Fluctuations in the radiation and thermal regimes of the atmosphere due to both back-scattering and absorption by aerosol particles has been considered in other studies. The second of these mechanisms can decrease the albedo of the earth-atmosphere system under certain conditions, which tends to increase the temperature at the earth's surface.

Quantitative criteria depending chiefly on the albedo of the earth's surface and determining the direction in which temperature changes under the influence of aerosol have been proposed in a number of studies. The values of these criteria found by different authors differ noticeably, which is explained both by particular features of the models used and by the fact that the values of the empirical coefficients characterizing back scattering and absorption of radiation by aerosol particles, did not coincide.

In some cases, causes for differences between these criteria are easily understood. For example, Mitchell suggested that the influence of aerosol on the thermal regime of the atmosphere is associated with a decrease in heat expenditure on evaporation from the earth's surface, due to a decrease in the influx of radiant energy. Mitchell apparently obtained the results by referring mainly to the weather conditions under which the condensation of water vapor is insignificant.

Let us consider results of one study along these lines (Kondrat'ev, et al., 1973), which concluded that the direction in which air temperature changes in the troposphere is determined by the ratio of the effective albedo of the aerosol component and the albedo of the earth's surface. If the first albedo is greater than the second, air temperature will decrease if aerosol is present, and vice-versa. A number of conclusions may be drawn from this work regarding the influence of tropospheric aerosol on the thermal regime for mean global conditions.

Calculations carried out in this study imply that the effective albedo of a "mean aerosol particle" varies from 0.20 to 0.60, depending on the coefficient of radiation absorption by aerosol particles and on the distribution of aerosol particles by dimensions. The mean effective albedo for the group of values given in this work was 0.42.

By comparing these values with typical values of the earth's surface albedo, we may conclude that under cloudless conditions atmospheric aerosol as a rule decreases air temperature over surfaces free of snow and ice, whose albedo is usually at most 0.20, and increases air temperature over a snow cover, whose albedo is usually greater than 0.60. Since the snow cover occupies a small part of the total surface of the globe, and since solar radiation in this zone is on the average less than its mean global value, it can be seen that a tendency predominates toward a decrease in mean air temperature in the presence of aerosol for that part of the earth's atmosphere free of clouds.

The influence of aerosol on mean air temperature when clouds are present is of significant interest. Kondrat'ev set the albedo of clouds equal to 0.70, which is a possible value for separate, thick cloud formations, but cannot be considered the mean global albedo of clouds.

The mean global albedo of clouds, according to data given in Chapter 2, is 0.46, which nearly coincides with the mean value of the effective aerosol albedo.

The difference between these two values is indisputably less than the probable error of each, which implies that the influence of aerosol on the thermal regime of the atmosphere is scarcely of substantial importance if clouds are present.

The same conclusion may be drawn by bearing in mind the fact that the bulk of tropospheric aerosol is situated below the upper cloud boundary, as a result of which the influence of aerosol on the radiation and thermal regime is substantially weakened if there are clouds.

We find that the combined influence of tropospheric aerosol on the mean air temperature is to decrease it by roughly half the amount that would be the case in a clear atmosphere, since on the average about half the earth's sky is covered with clouds.

It is difficult to quantitatively calculate the influence of tropospheric aerosol on the thermal regime of the atmosphere, since not enough reliable data are available for mean climatological parameters in the models that are used as, for example, the coefficients of radiation absorption by aerosol particles. Thus it is of substantial importance to analyze empirical data in order to estimate the influence of tropospheric aerosol on the radiation and thermal regimes.

Observational data from a group of actinometric stations situated in Europe, Asia, and America were processed by Z. I. Pivovarova and used by Budyko and Vinnikov (1973) to estimate the influence of anthropogenic aerosol on climate. A graph describing the secular variations of direct solar radiation over the past several decades in the case of a clear sky was constructed from these data.

Figure 31 presents values of direct radiation anomalies expressed in

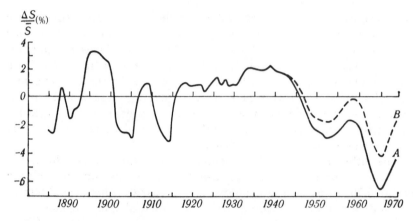

Fig. 31. Secular variations of direct radiation.

percentages of its norm from 1885 to 1970. Curve A in Fig. 31 characterizes values averaged over 5-year running periods. (The use of 5-year averaging leads to a number of differences between this curve and those presented above, which were averaged over 10-year periods.) It should be noted that the number of stations at the end of the nineteenth and beginning of the twentieth centuries was less than during the recent period. Therefore values for that interval of time are less reliable.

It is evident from Fig. 31 that several sharp decreases in the magnitude of direct radiation occurred from the end of the nineteenth century to the middle of the 1910s as a result of powerful volcanic eruptions, such as Krakatau, Mont Pelée, Katmai, and others, after which the aerosol concentration in the lower layers of the stratosphere significantly increased. Direct radiation attained maximal values and varied comparatively little from the end of the 1910s to the beginning of the 1940s, after which it began to gradually decrease.

A number of studies (McCormick and Ludwig, 1967; Bryson, 1968; Budyko, 1969; Man's Impact on the Global Environment, 1970; Inadvertent Climate Modification, 1971) have claimed that the gradual decrease in direct radiation beginning with the 1940s can be explained by the increase in the mass of atmospheric aerosol created by human activity.

We noted above that anthropogenic aerosol is basically concentrated in the lower layers of the troposphere. Its influence on the radiation that reaches the earth's surface is therefore significantly less under cloudy conditions than with a clear sky. We also said that definite relationships exist between the variations in direct radiation as a result of scattering on aerosol particles and the corresponding change in total radiation. Using these relationships, we may calculate the variations in total radiation from data on fluctuations of direct radiation.

If these fluctuations result from radiation that has scattered on aerosol particles in the stratosphere, the corresponding calculation can be performed by using data on variations of direct radiation represented by curve A in Fig. 31. But if part of these fluctuations is due to attenuation of radiation in the lower layers of the troposphere, the influence of clouds on the radiation regime of the atmosphere must be taken into account to calculate the mean global variations in total radiation.

Let us assume that since the beginning of the 1940s direct radiation has varied because of two processes, namely the growth in the mass of anthropogenic aerosol, which would monotonically decrease the amount of radiation, and a fluctuation in the transparency of the stratosphere, which would decrease direct radiation as aerosol concentration in the stratosphere increased, or increase the direct radiation to a limit corresponding to total or nearly total absence of aerosol in the stratosphere.

We will also assume that the variation in direct radiation due to an increase in anthropogenic aerosol is characterized by a straight line passing through points of the secular variation curve corresponding to atmospheric transparency maxima in different years. We may suppose that the stratosphere was either completely or nearly completely free of aerosol particles with increased transparency of the atmosphere, which is confirmed by data of direct observations of stratospheric aerosol obtained by Cadle (1972) and by other investigators. The estimate found here for the decrease in radiation by anthropogenic aerosol corresponded to clear sky conditions. The mean cloud cover for the globe was taken to be 0.5 and the influence of anthropogenic aerosol on the total radiation in the case of a cloudy sky is considered small in comparison with its influence if no clouds are present. We then find that values of direct radiation must be increased by one-half of the variation in direct radiation for a clear sky caused by anthropogenic aerosol, in order to calculate the influence of radiation changes on air temperature. The secular variation of direct radiation, constructed by taking into account this correction, is depicted in Fig. 31 in the form of the broken line B.

It is possible to approximate the mass of anthropogenic aerosol from variations in direct radiation resulting from an increase in this mass. It is evident from Fig. 31 that anthropogenic aerosol has recently decreased the direct radiation in clear sky by about 6%. This corresponds to a decrease in total radiation by roughly 1% for average conditions in the case of a clear sky. According to the data given in Chapter 3, roughly $10^6 \, g/cm^2$ of aerosol will change the total radiation by 1% as a result of scattering by aerosol particles. Thus the amount of anthropogenic aerosol is about 5 million tons for the entire globe.

This amount corresponds to the mass of optically active aerosol with particle dimensions of tenths of a micrometer. It is usually assumed that approximately one-half the mass of aerosol in the troposphere consists of so-called gigantic particles that are insignificantly smaller in number than particles with dimensions of tenths of a micrometer. As a result the gigantic particles do not substantially induce any attenuation of radiation (Junge, 1963). Taking this into account, we find that the total mass of anthropogenic aerosol can reach about 10 million tons.

Let us compare this value to results obtained by direct methods. According to modern data (Man's Impact on the Global Environment, 1970), between 1,000 and 2,600 million tons of matter forming aerosol particles annually enter the troposphere, 20% of this amount being created by human activity. If we assume that the mean lifetime of particles in the troposphere is roughly 10 days (Inadvertent Climate Modification, 1971), we find that the total aerosol mass in the troposphere is equal to 30–70 million tons, while the mass of anthropogenic aerosol is about 3–14 million tons. The above estimate for the amount of anthropogenic aerosol reasonably agrees with these values. This confirms that the influence of human activity on direct radiation is calculated with sufficient accuracy.

Fluctuations in the mean latitude air temperatures from 1910 to 1950 were calculated in Chapter 3 from data on fluctuations in solar radiation. It is necessary to take into account the influence of human activity on climate in order to explain the causes of more recent changes in air temperature.

We demonstrated above that over recent decades variations in direct radiation in the case of a clear sky have depended not only on fluctuations in aerosol concentration in the stratosphere, but also on variations in the amount of aerosol in the troposphere resulting from human activity. It can be seen from the data presented in Fig. 31 that variations in the amount of tropospheric aerosol have had a gradually increasing influence on radiation and now exert a greater influence than variations of the aerosol mass in the stratosphere.

Increase in the concentration of carbon dioxide in the atmosphere has, in addition, influenced the air temperature of the earth's surface over recent decades. For this reason, the mean temperature of the earth's surface should have grown over recent decades by approximately 0.3–0.4°. Since, according to observational data, the mean air temperature in the northern hemisphere has decreased by roughly 0.3° within this period of time, we may question whether an increase in the mass of anthropogenic aerosol could decrease the mean temperature of the earth's surface by a value consistent with these measurements.

This question can be answered only by estimating the influence of variations in the aerosol concentration in the troposphere on air temperature at the earth's surface, which is more complicated than the analogous influence of variations in aerosol concentration in the stratosphere (Budyko, 1974a).

The data presented above imply that an increase in the concentration of anthropogenic aerosol decreased direct radiation in the northern hemisphere by roughly 6%. In the case of a clear sky, the level of radiation would then decrease by approximately 1%. Since tropospheric aerosol does not strongly affect the total radiation if clouds are present, and also since the mean planetary cloudiness is roughly 0.5, we find that the mean decrease in total radiation is 0.5%. By using the semiempirical theory of the atmospheric thermal regime, we find that such a decrease in total radiation would lead to a decrease of 0.75° in the mean temperature of the earth's surface. This result corresponds to the case in which aerosol particles only scatter radiation, but do not absorb it. A number of studies have indicated that absorption of short-wave radiation by tropospheric aerosol particles can compensate to some degree the temperature drop resulting from back scattering of radiation by aerosol particles.

It is difficult to estimate the influence of radiation absorption by aerosol due to the absence of sufficiently general studies of the mean values of absorption coefficients. It is, nevertheless, possible to use results of a study by Yamamoto and Tanaka (1972), in which the drop in the mean global temperature as a result of radiation scattering by aerosol with and without radiation absorption was calculated using a semiempirical theory of the thermal regime. Yamamoto and Tanaka obtained a ratio of 0.68 between the temperature fluctuation in the first case and the temperature fluctuation in the second case with a mean global cloud cover by assuming that atmospheric opacity was not too high. If we multiply this value by the temperature drop determined above, we find that this drop is nearly 0.5° for actual conditions.

Though our calculation is highly approximate, it nevertheless gives us grounds for assuming that changes in the mean global air temperature over recent decades may partly be explained in terms of man's economic activity.

This confirms previous conclusions obtained in a number of the studies mentioned above, as well as by Bryson (1972).

Figure 31 shows that after the mid-1960s direct radiation stopped decreasing. This fact is worthy of notice because it possibly points to a stabilization of the aerosol mass that is of considerable significance for estimating the climatic conditions of the last decade.

The influence of aerosol particles on direct radiation determined by actinometric observations in Europe, Asia, and North America was used in this discussion as more or less corresponding to the mean conditions of the entire northern hemisphere.

This hypothesis is strongly justified in the case of stratospheric aerosol whose concentration varies comparatively slightly within the hemisphere. It is less obvious for tropospheric aerosol due to its limited lifetime in the troposphere. Observational data indicate that extensive regions can be found in the lower layer of the atmosphere with significantly increased aerosol particle content.

Thus the extent to which actinometric data used in this analysis are representative enough for the hemisphere as a whole requires further study.

Other Factors of Climatic Changes. It is likely that influence of all other anthropogenic factors than the quantity of atmospheric carbon dioxide and aerosol, on the contemporary global climatic conditions is comparatively small.

One of these factors involves the production of energy in the course of economic activity, which leads to additional heating of the atmosphere and of the earth's surface. All energy used by man is in the end converted into heat, the chief part of it being an additional source of energy for the earth that increases its temperature.

Hydroelectric power and the energy in wood and agricultural products are the only substantial components in man's modern use of energy that involve transformation of solar energy annually absorbed by the earth. Consumption of these types of energy does not vary the earth's heat balance and does not lead to additional heating.

However, these types of energy constitute a small part of all the energy used by man. Other types of energy, such as the energy of coal, petroleum, natural gas, as well as atomic energy, are new sources of heat that do not depend on transformation of solar radiation in the contemporary epoch.

In previous works (Budyko, 1962) we estimated the amount of heat generated in the course of human economic activity. This amount is not great, being about 0.01 $kcal/cm^2 \cdot yr$ for the earth as a whole. It is about 100 times greater, and reaches 1–2 $kcal/cm^2 \cdot yr$, over the most highly developed industrial regions that extend over tens and hundreds of thousands of square kilometers. It grows to tens and hundreds of $kcal/cm^2 \cdot yr$ over large cities (tens of square kilometers). Let us estimate how this additional production of heat influences the mean temperature of the earth.

We noted above that a 1% change in the influx of solar energy will change the mean temperature of the earth's surface by 1.5°. Since the production of heat as a result of human activity constitutes about 0.006% of the total amount of radiation absorbed by the earth-atmosphere system, the increase in mean temperature corresponding to this amount of heat is roughly 0.01°. This value is comparatively small, though the increase in temperature can be significantly greater in different regions since heat sources created by man are distributed highly nonuniformly on the earth's surface.

It was also noted in this work that temperature could increase about 1° in the most developed industrial regions if no atmospheric circulation is present, and by tens of degrees in large cities, which would obviously make life there impossible. The influence of atmospheric circulation noticeably weakens corresponding temperature increases, and this weakening is greater, the less is the area over which the production of the additional thermal energy is concentrated.

A detailed analysis of the energy consumption that represents an additional heat source for the atmosphere was carried out by Flohn (Inadvertent Climate Modification, 1971). He found that the heat influx created by man in the central part of New York and Moscow is several times that of the amount of energy arriving from the sun. The influx of additional heat is between 10 and 100% of the influx of solar energy in a number of smaller cities as well as in the most highly developed industrial regions tens of thousands of square kilometers in area.

The additional influx of heat is 1% of the energy of solar radiation received in many countries measuring up to hundreds of thousands of square kilometers in area.

Flohn's data confirm our previous conclusion that man's production of energy is a substantial climate-forming factor not only in large cities, but also over significant areas of industrial regions.

In the preceding section it was noted that irrigation of desert areas exerts a definite influence on the mean temperature at the earth's surface. Let us determine how a currently existing irrigation system affects the mean temperature of the earth's surface. The albedo of the earth's surface may decrease by about 0.10 as a result of irrigating arid regions. By taking into account the relation between the albedo of the earth's surface and the albedo of the earth-atmosphere system (cf. Chapter 2), we find that with scarce cloudiness this decrease in the albedo of the earth's surface corresponds to a 0.07 decrease in the albedo of the earth-atmosphere system.

There are now roughly two million square kilometers under irrigation, which constitute about 0.4% of the total surface of the earth. Thus irrigation decreases the earth's albedo by roughly 0.03%.

It was established in Chapter 2 that a variation in the earth's albedo by 1% will vary the mean temperature of the earth's surface by 2.3°. With this in mind, we find that irrigation increases the mean temperature of the earth's surface by approximately 0.07°. This temperature fluctuation is not negligibly small and may play an important role if the areas under irrigation are increased.

The construction of reservoirs can exert, in conjunction with irrigation, a special influence on the mean temperature of the earth's surface.

The mean value of the albedo of the earth's surface decreases when reservoirs are constructed in planted regions by approximately the same amount as when desert regions are irrigated. However, since the largest artificial reservoirs have been built in regions with a comparatively moist climate where a more or less dense cloud cover exists, the albedo of the earth-atmosphere system in this case varies less than over irrigated regions, where cloudiness is low. In addition, since the total area of artificial reservoirs is appreciably less than the area of irrigated land, their influence on the mean temperature of the earth's surface is comparatively slight.

Release of heat by people in the course of their activity is one more process that may exert a strong influence on air temperature. This heat influx is significant only over limited spaces, where large crowds are concentrated, and its role is slight in the case of larger areas.

The data of this section demonstrate that changes in the global climate over the last several decades have apparently depended heavily on man's economic activity.

This may explain why the relations between fluctuations in mean global temperature and volcanic activity (cf. Chapter 3) established from earlier observational data have been upset in this era.

The further evolution of man's economic activity will lead to even more changes in the global climate than have occurred till now.

7 The Climate of the Future

7.1 The Outlook for Climatic Changes

Natural Climatic Changes. In the study of future climatic changes, we must first discuss those changes that may result from natural causes.

Chapters 3 and 4 considered the influence of a number of factors on climatic changes, each of those factors being characterized by different time scales. These include the following fundamental factors.

1. *Volcanic activity.* A study of contemporary climatic changes shows that fluctuations in volcanic activity can affect climatic conditions for years and decades. Volcanism can also influence climatic changes over periods on the order of centuries and over longer intervals of time.

2. *Astronomical factors.* A change in the position of the earth's surface relative to the sun will create climatic changes with time scales of tens of thousands of years.

3. *Composition of the atmosphere.* A decrease in the content of carbon dioxide in the atmosphere particularly influenced the climate towards the end of the Tertiary and Quaternary periods. Bearing in mind the rate of this decrease and the air temperature changes corresponding to it, we may conclude that the influence of natural variations in carbon dioxide content is of importance for intervals of time greater than 100,000 years.

4. *Structure of the earth's surface.* A variation in topography and related variations in the position of the shores of seas and oceans change climatic conditions over large spaces for periods of time of at least hundreds of thousands and millions of years.

5. *Solar constant.* We should bear in mind that slow fluctuations of solar radiation are possible as a result of the evolution of the sun, setting aside the question of the existence of short-period fluctuations in the solar constant. These fluctuations can substantially influence climatic conditions over periods of time lasting at least 100 million years.

Climatic conditions can vary as a result of self-oscillating processes in the atmosphere-ocean-polar ice system appearing alongside external factors. These variations have periods of years and decades and, possibly, hundreds and even thousands of years.

The time scales in this list for the action of different factors on climatic changes basically agree with similar estimates made by Mitchell (1968) and others.

Let us consider how these factors can influence the climatic conditions of the future.

No reliable methods for predicting the level of volcanic activity are available at the present time. Thus, in studying the influence of this factor on climate, we are forced to limit ourselves to estimating its probable scale. It was indicated in Chapter 3 that the rise in temperature in the first half of the twentieth century, due basically to fluctuations in volcanic activity, increased the mean global temperature by roughly 0.5°. It is evident from an earlier study (Lamb, 1970) that the abating in volcanic activity in the 1920s and 1930s constituted a large and comparatively rare anomaly in volcanic activity over recent centuries.

A quantitative analysis of data from Lamb's report on volcanic activity as well as data on the secular variations of mean global temperature lead us to conclude that the mean-decade global temperature will probably change toward the end of the current century due to fluctuations in volcanic activity by about 0.2°, i.e., by an amount smaller than the below estimate of the probable change in air temperature due to man's economic activity.

Unlike predictions of the level of volcanic activity, future changes in the position of the earth's surface relative to the sun can be calculated with reliable accuracy. Such calculations (Sharaf and Budnikova, 1969; Vernekar, 1972) allow us to estimate fluctuations in the influx of solar energy over the warm part of the year, the position of the boundaries of the polar ice cap strongly depending on these fluctuations. Milankovich (1930) noted that fluctuation of radiation near the polar circles exert a particularly strong influence on the movement of ice.

The data of these calculations imply that radiation will sharply decrease (by two-thirds the decrease of the last Würm glaciation) 10,000–15,000 years within the "critical latitudes" of the northern hemisphere. This decrease in radiation can lead to the onset of a new ice age with continental glaciations somewhat less than the glaciations of the Würm period.

We should later expect a marked increase in radiation, which could lead to destruction of the glaciers, after which decreases in radiation in the critical latitudes of the northern hemisphere will repeat in about 50,000 and 90,000 years. The amplitude of these radiation decreases will grow, approaching the magnitude of the radiation decrease that brought about the last Würm glaciation.

Much greater radiation decreases will occur in the critical latitudes of the northern hemisphere in several hundreds of thousands of years, which indicates that immense glaciation may develop in these epochs.

The influence of astronomical factors on the climate of the future may significantly intensify with a further decrease in the content of carbon dioxide in the atmosphere. Since such a trend has prevailed for many millions of years, it is highly likely that the amount of atmospheric carbon dioxide will continue to decrease in the future without man's economic activity. The rate of this decrease can be approximated from data given in Chapter 4 on its decrease over the last geological epoch, the Pleistocene.

Extrapolation can be used to estimate the period of time over which the carbon dioxide concentration of the atmosphere will decrease from the value it had prior to the epoch of rapid industrialization (0.029%) to a value at which total glaciation of the planet (0.015%) is possible. The difference between these two values is close to the change in carbon dioxide concentration that has come about since the beginning of the development of Quaternary glaciations through the contemporary epoch. Thus, we may conclude that total glaciation of the earth can occur in one million years.

This period must decrease if we bear in mind that the general tendency toward the development of glaciation resulting from a decrease in the concentration of carbon dioxide is accompanied by a periodic increase of ice covers due to a variation in the position of the earth's surface relative to the sun. Thus total glaciation of the earth will probably begin in several hundreds of thousands of years. This estimate can be used to develop a picture of the possible evolution of the biosphere.

The continuing decrease in carbon dioxide concentration will be accompanied by a gradual drop in the productivity of autotrophic plants and a lower total mass of living organisms on the earth. The zone of polar glaciation will simultaneously gradually expand, migrating to lower latitudes as the ice ages begin.

The productivity of autotrophic plants and the total volume of biomass on the earth will decrease by at most one-half when the ice cover reaches the critical latitude. The ice cover will then spread to the equator as it spontaneously develops. As a result, total glaciation of the planet and a very stable glaciation due to the low negative temperatures over the entire globe will set in.

Total glaciation could lead to the cessation of all biological processes on the planet. This hypothesis is based on the fact that no living organism that could permanently exist in the central regions of an Antarctic glacier has appeared during the long history of the glaciers. It would be difficult to expect any living organisms to arise or to adapt to such unfavorable conditions as the climatic conditions of Central Antarctica spread over the entire globe within a comparatively short period of time.

Total glaciation of the planet corresponding to the temperature conditions indicated in Fig. 8 by point B' will be extraordinarily stable, since air temperature at all latitudes will be much below freezing point. However, it is possible that such glaciation can vanish in the distant future as a result of a gradual increase in the solar constant. A number of astronomical studies have expressed the hypothesis that the temperature of the sun's surface and the amount of energy radiated by it may increase in the course of its evolution (Schwarzschild, 1958; Ängstrom, 1965; White, 1967).

It is evident from Fig. 8 that an increase in the solar constant by roughly 40%, which can occur, according to these works, within several billion years, is required to upset planetary glaciation. In this case, ice will melt at all latitudes and the air temperature will increase by more than 100°, reaching a mean value of about 80°C. Since the temperature of the earth's surface will continue to rise with a further increase in the solar constant, the melting of ice will hardly lead to conditions favorable for the reappearance of life on earth.

This path of development of the earth's thermal regime is extremely hypothetical, since it is not entirely clear how the sun has evolved. It is, however, of some interest that if this variation holds, the thermal regime will pass through several stages corresponding to the large segment of the hysteresis curve in Fig. 8, i.e., from the ice-free conditions of the Mesozoic (somewhat above point E) to contemporary conditions (point 3), to total glaciation (from point B' to A) and to the destruction of glaciation (point A'). This estimate of time within which total glaciation of the planet can occur is very approximate. The accuracy of the calculation is substantially limited by the method used for determining the rate at which the carbon dioxide content decreases in the atmosphere, which is highly tentative. This is in addition to the highly schematic nature of the model describing the thermal regime of the atmosphere that was used to obtain this estimate.

7 THE CLIMATE OF THE FUTURE

This result is of interest chiefly as an illustration of a previously expressed conception of the specific nature of Quaternary glaciations (Budyko, 1971). It seems possible to us that, unlike the Permo-Carboniferous and other ancient glaciations, Quaternary glaciations were not temporary episodes in the evolution of the earth, but represented the beginning of a transition from stable ice-free climatic conditions to even more stable total-glaciation conditions of the planet. The duration of this transitional period in which the existence of the biosphere may conclude is much smaller than the total duration of life on our planet, according to the estimate presented above.

The probability of total glaciation in the course of the earth's natural evolution increases if we take into account existing tendencies toward an increase in the elevation of the continents in a number of high- and middle-latitude regions of the northern hemisphere.

Saks (1960) and other authors have presented paleogeographic maps showing that by the end of the Jurassic the oceans occupied nearly half the latitude circle at 70°N, which promoted free circulation of sea water between the middle and polar latitudes. At the beginning of the Cretaceous, the areas occupied by the continents began to gradually increase at this latitude, and this has continued in our time, as a result of which the continents now occupy five-sixths of this latitude circle.

The width of the straits connecting seas of middle and high latitudes in the northern hemisphere was reduced to one-third. There was thus a marked decrease in meridional heat transport by sea currents to polar latitudes, and the prerequisites for the development of arctic glaciation were created.

There are no grounds for believing that the near future will see an end to the increasing isolation of the northern polar basin, a trend that has already lasted many millions of years. If this trend continues, it will become more likely for the glaciers to reach the critical latitude in the future.

The influence of changes in the earth's surface on the climate of the future will probably manifest itself after a decrease in the carbon dioxide content in the atmosphere has set in and may therefore turn out to be insignificant. However, since the corresponding estimates are highly approximate, the comparative importance of these two factors for the climatic conditions of the future is not entirely clear.

A hypothetical scheme for future variations of the mean global air temperature at the earth's surface in the light of these hypotheses is presented in Fig. 32, which also depict fluctuations in the mean global temperature in the past.

For the sake of clarity, the scale of the time axis in Fig. 32 for the

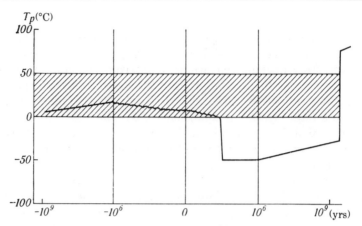

Fig. 32. Variations of the mean air temperature at the earth's surface.

previous and next million years is 10^3 times greater than for more distant times. The range of mean temperatures which are most favorable for biological processes is shown by the hatched area. We should emphasize that this scheme is highly tentative in the light of the hypotheses assumed in constructing it.

Inadvertent Climate Modification. The development of civilization, an exceptionally rapid event on the time scale of geological history, has radically changed the outlook for the further existence of the biosphere.

We need only note that the concentration of carbon dioxide in the atmosphere has increased by 0.003%, i.e., by 10% of its magnitude over recent decades, as a result of the burning of different types of fuel. This increase compensates for its decrease over the previous 200,000 years. Thus, human activity has changed the direction of the process by which the concentration of atmospheric carbon dioxide has varied and has increased its rate by a few thousand times.

Though in this case man's effect on climate has been unintentional, it has nevertheless already become of great importance for preventing the further development of glaciation.

Let us imagine the highly unlikely case that man's effect on the atmosphere may cease in the future. Under these conditions, the increase in carbon dioxide concentration in the atmosphere that has been attained over the last century will defer planetary glaciation by thousands of years. Obviously, global glaciation would no longer be possible if the contemporary rate of man's effect on the atmosphere is maintained and, especially, if it increases.

In addition, the rapidly increasing influence of man's economic activity on climate will make possible in the comparatively near future a significant change in the global climate.

This change, of course, will not be as catastrophic as a total planetary glaciation, though it may create a number of substantial difficulties for further economic development. The outlook for anthropogenic climatic changes requires great attention.

Let us consider the influence of three fundamental factors on the climatic conditions of the next century:

1) a growth in the production of energy consumed by man;
2) an increase in carbon dioxide content in the atmosphere;
3) fluctuations in the concentration of atmospheric aerosol.

We noted in our first work dealing with this question (Budyko, 1962) that a 4 to 10% increase in the annual production of energy can equalize the heat created by man with the radiation balance of the entire surface of the continents within 100–200 years. It is obvious that enormous climatic changes could then be expected over the entire planet.

The data presented in Fig. 33 provide some idea of the possible influence of a growth in energy production on the climate of the future (Budyko, 1972). Here, curve 1 represents the secular variations of a deviation of mean air temperature from the norm for the northern hemisphere, based on observational data. Curve 2 gives the results of calculations of changes in mean planetary temperature with an increase in energy production of 6% per year. This calculation is based on the use of our previous model under the assumption that a temperature change occurs as a result of a variation in heat influx, all other factors influencing the climate, including the albedo of the earth-atmosphere system, being kept constant. As we have shown above, this hypothesis does not take into account the possibility that polar ice could melt, which leads to an underestimate of future temperature increases.

It is evident from Fig. 33 that the rise in temperature due to a growth in energy production during the first half of the twenty-first century will exceed the temperature changes that occurred in the first half of the twentieth century as a result of natural causes. Temperature will subsequently grow rapidly, which will lead to great changes in global climate. Thus a growth in the production of energy consumed by man may exert a substantial influence on the climate of the future.

The second factor that may significantly change the climate is the growth in the carbon dioxide concentration in the atmosphere. Observational data indicate that it has been mounting yearly at a rate of about 0.2% (of its total amount) over the most recent decades because of the

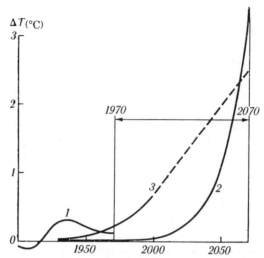

Fig. 33. Secular variations of air temperature anomalies at the earth's surface. (1) Variations of mean temperature; (2) variations due to growth of energy production; (3) variations due to growth in carbon dioxide concentration.

burning of a growing amount of fuel. This process may have accelerated somewhat over recent years. On the basis of these data, Machta and other authors have concluded that the carbon dioxide concentration in the atmosphere will increase by 15–20% between 1970 and 2000 (Inadvertent Climate Modification, 1971).

We indicated in the preceding chapters that an increase in the amount of atmospheric carbon dioxide will lead to a change in the radiation regime of the atmosphere. The transparency of the atmosphere to long-wave radiation will decrease, and this will increase the air temperature at the earth's surface. Assuming such an increase in carbon dioxide concentration, Manabe (1970) found that (neglecting fluctuations in the area of the snow-ice cover) the mean planetary air temperature at the earth's surface in the year 2000 may be approximately 0.5° higher, and the mean temperature at high latitudes approximately 1° higher than in 1970. Similar results are obtained by applying the dependence, given in Chapter 4, of air temperature on carbon dioxide concentration with constant earth-atmosphere albedo. A change in global temperature corresponding to this calculation is shown in Fig. 33 by the unbroken segment of curve 3.

It is evident from Fig. 33 that by the year 2000 a change in temperature due to the increased carbon dioxide concentration will exceed the anomalies in the natural fluctuations of the global temperature observed during the first half of the twentieth century. If the carbon dioxide concentration continues to grow in the twenty-first century, more telling climatic changes may ensue.

The change in the concentration of atmospheric aerosol is an additional factor resulting from man's economic activity that may lead to climatic fluctuations. Data were presented in Chapter 6 implying that this increase affects the contemporary climate. Obviously a further growth in aerosol concentration can entail a sharp climatic change. It is, however, rather difficult to predict the amount of atmospheric aerosol in the future.

It is worth noting that the latitudinal distribution of absolute values of temperature variations are similar with increasing carbon dioxide as well as with aerosol concentration. Since these changes are of opposite sign, we may suppose that in particular cases they compensate each other to some degree.

Contemporary anthropogenic climate changes, which have somewhat abated as a result of the counterinfluence of changes in the concentration of atmospheric carbon dioxide and aerosol on the thermal regime, constitute one example of such counterbalancing.

It has been proposed in the author's previous work (Budyko, 1972b) that, when arrangements to protect the atmosphere against its pollution by man's activity are realized, the growth of the anthropogenic aerosol mass has to stop. In this connection it was assumed probable that future climatic changes would be considerably dependent on the influence of the growth of carbon dioxide concentration on the atmosphere thermal regime, which would soon lead to developing the new warming (according to the curve 3 in Fig. 33).

It was noted in the above work that the new warming could be of great practical importance due to deterioration of precipitation conditions over much of the continents in moderate latitudes (that will result in droughts) and due to decreasing the area of polar sea ice.

In passing to the possible climatic conditions of the future, let us note that the practical value of this question is a function of the scale of possible climatic changes and the time at which they may begin. If these changes will be sufficiently great and will occur in the not too distant future, the prediction of future climatic changes clearly becomes one of the most important problems of contemporary meteorology.

The importance of this problem is particularly evident in the strong dependence of the national economy of all countries on contemporary climatic conditons. A sharp climatic change requires enormous capital investment to accommodate economic activity to these new conditions.

The time by which it is necessary to have information on climatic changes is apparently comparable with the likely lifetime of currently planned industrial and agricultural buildings and systems whose functioning is subject to climatic conditions. This period is at least 100 years for the most durable buildings. If climate will substantially change in the future, this possibility must obviously be taken into account to some degree in planning these buildings.

The time required to prepare and carry out measures to control climatic changes and adjust the national economy to these changes is a second criterion that can be used to estimate the time by which it is desirable to have information on possible climatic changes. This time is probably at least several decades, since such measures require the solution of many complex scientific and technological problems in order to be implemented. Thus, it is desirable to have information on changes in climatic conditions that may occur within 100 years in the future.

The problem of predicting climatic changes as a result of human activity substantially differs from weather forecasting. While we may limit ourselves in the latter case to an analysis of the physical processes in the atmosphere and hydrosphere, the former problem requires that we also bear in mind the changes that time may bring to the indices of man's economic activity.

Thus the problem of predicting climatic changes involves two fundamental elements: predicting the development of a number of aspects of economic activity (growth in the use of fuel, increasing concentration of carbon dioxide in the atmosphere, growth in the production of energy, etc.) and the evaluation of those climatic changes that may change the corresponding indices of man's activity.

This contributes to two important features involved in these predictions. On the one hand, they will be unavoidably provisional. Man's economic activity is not independent of his influence on climatic conditions. In particular, if this activity leads to extremely unfavorable climatic changes, it is highly likely that it will have altered its character before these changes begin. Therefore the problem of a climatologist is not to predict the actual climate of the future, but to calculate the parameters of this climate for a number of possible forms of economic activity. We may optimize long-range planning for the development of the national economy, taking measures against unfavorable changes and taking into account results of such a calculation. Thus prediction of possible climatic changes may serve as a basis for taking measures to modify climate.

The second feature of predictions of the future climate involves their probable error. Since it is difficult for a number of reasons to quantitatively predict economic activity for several decades in advance, the precision of such a prediction cannot be high. Thus we are apparently justified in using schematic models of climate theory, which, however, must yield a correct estimate of the direction and order of magnitude of possible climatic changes.

The practical importance of such estimates lies in the fact that they allow us to identify those forms of economic activity that can lead to great (i.e., the most consequential to the national economy) climatic changes exceeding the error of the corresponding calculations.

We must first discuss the effect of a further growth in energy production, which we spoke of above, in order to deal with the possible climatic changes in the next century.

It is evident from Fig. 33 that even disregarding the feedback between the polar ice regime and air temperature, we may conclude that a rapid increase in planetary temperature will begin in the middle of the twenty-first century if energy production annually grows by 6%. This increase will be accompanied by enormous climatic changes, which may lead to catastrophic consequences for the national economies of many countries. Thus uncontrolled growth in heat production will lead to a so-called "heat barrier" in the development of energetics.

We should note that the assumptions made in deriving curve 3 in Fig. 33 are not entirely unrealistic. A 6% annual increase in energy production has been the case over recent years (Inadvertent Climate Modification, 1971). The magnitude of the additional heat influx per unit area reached by 2070 obtained in this calculation corresponds to the amount of heat now released in some regions of the industrially most developed countries.

A review of a number of works in which the possible use of energy in the future is estimated can be found in an article by Kellogg (1974). These data imply it is highly likely that the global use of energy will increase more than 100-fold in the comparatively near future. Such an increase will significantly change climatic conditions, particularly as a comparatively small increase in the air temperature begins to affect the polar ice, whose recession will lead to an additional sharp temperature change at high latitudes.

Thus the possibility that a heat barrier will be reached within a period of at most 100 years must be considered one of the fundamental problems that will confront the technology and energetics of the not too distant future.

The methods to solve it require special discussion outside the scope of this book. Here we may only recall N. N. Semenov's advocacy of widespread use in the future of solar energy for economic purposes in order to limit overheating of the earth's atmosphere.

The state of the ice cover in high latitudes is of great importance in studying climatic changes that can be attributed to an increase in global temperature. In our previous works (Budyko, 1962a, 1971) we offered a number of methods to estimate the influence of changes in meteorological factors on polar ice. One such method, based on a calculation of the heat balance of the ice cover, allowed us to calculate a change in ice thickness for a given air temperature increase. This demonstrated that polar sea ice will completely melt in several years if the air temperature in the warm part of the year increases by 4°C.

Since the feedback between the area of the ice cover and the air temperature was not taken into account in this calculation, it is entirely evident that the magnitude of the temperature anomaly sufficient for complete melting of ice is overstated. Nevertheless, we may conclude that the temperature increase which will be reached in roughly 100 years according to the data of Fig. 33 will lead to total thawing of polar sea ice.

Obviously, sea ice may partially melt before a heat barrier is reached, i.e., before the mean global temperature increases by several degrees.

Let us use equation (3.1) to estimate changes in the area of polar ice. We will assume that the magnitude of the relative variation in global radiation in this formula can be expressed in terms of the change in global temperature corresponding to it, which is determined in accordance with the semiempirical theory of the thermal regime with constant albedo. In this case, the area of polar ice can be calculated for different values of a mean global temperature increasing as a result of an increase in carbon dioxide concentration and energy production in accordance with the data depicted in Fig. 33.

Since the influence of energy production on the thermal regime of the atmosphere will be comparatively slight prior to 2000, the air temperature of the earth's surface will be chiefly determined by the growth in carbon dioxide concentration if the mass of atmospheric aerosol remains constant.

The mean global air temperature in the local Manabe-Wetherald model (1967) will increase by 0.5° if carbon dioxide concentration increases by 20%, and it will increase by 0.7° according to a three-dimensional model of climate theory that takes into account the influence of the snow cover on the thermal regime (Smagorinsky, 1974).

A similar calculation using our semiempirical model of the atmospheric thermal regime under the hypothesis that the polar ice is quasi-stationary yields a temperature increase of 0.6°.

All these models imply that an increase in the mean annual air temperature at high latitudes is roughly twice as great as at low latitudes.

Results of a calculation of ice boundary changes are shown in Fig. 34, where the mean latitude of the sea ice boundary in the northern hemisphere is laid out along the vertical axis (Budyko, 1972b).

It is evident from Fig. 34 that the mean boundary of polar ice will recede roughly 2° to the north by the year 2000. This change in the boundary of polar ice corresponds to a much greater decrease in its area than occurred during the warming period in the first half of the twentieth century and also to a stronger increase in mean air temperature than its level at the end of the last century. A decrease in the mean area of polar ice will exert a noticeable influence on air temperature at high latitudes.

It is necessary to conjecture what might be the carbon dioxide concentration in the atmosphere for the twenty-first century in order to estimate further changes in the boundary of polar ice. It was assumed in the calculations that the mean air temperature will increase linearly under the influence of a growth in carbon dioxide concentration after the year 2000. Due to the rapid growth in energy production in the twenty-first century, the assumption that carbon dioxide concentration will change in this century will not greatly influence the results of the calculations.

Errors due to the inaccuracy of equation (3.1) for great changes in the area of ice may exert a more substantial influence on the computation of the time necessary for complete thawing of polar ice. As we noted above, it is not precluded that polar ice is unstable when comparatively

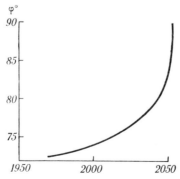

Fig. 34. Variations of polar sea ice boundary in the northern hemisphere.

small in area and can thaw without any additional increase in the temperatue of the lower air layers. In such a case, the data presented in Fig. 34 will overstate the period of time necessary for complete thawing of ice.

It is evident from Fig. 34 that under these hypotheses the time necessary for complete thawing of sea ice in the Arctic is roughly 80 years.

That this may be an overestimate follows from a calculation of the increase in mean global temperature corresponding to the moment at which sea ice has completely thawed. This value is approximately 3.5°C, i.e., close to the value presented above, which insures thawing of polar ice without taking into account the feedback of a change in the ice cover on the thermal regime.

A number of indices describing changes in climatic conditions associated with the thawing of polar ice are shown in Fig. 35. Temperature changes in middle and high latitudes of the northern hemisphere by the year 2050, i.e., when thawing of the polar ice will end, are depicted in Fig. 35 as calculated from the equations of the semiempirical theory of the thermal regime. It is evident from this figure that complete thawing of polar ice will lead to enormous changes in the thermal regime in the high latitudes.

A comparison of the data of Fig. 35 with paleoclimatic data demonstrates that, assuming ice-free conditions in the Arctic, the thermal regime in middle and high latitudes of the northern hemisphere will be similar in many respects to the regime at the end of the Tertiary, millions of years ago. Thus the transition to a new thermal regime can be represented as a return to the climatic conditions of the pre-Quater-

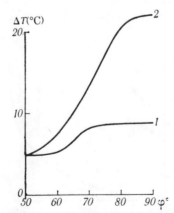

Fig. 35. Changes in mean latitudinal temperature with thawing of polar ice. (1) Warm half-year; (2) cold half-year.

nary. The rate of this return will be enormous – roughly 10^5 greater than the rate of natural cooling over the recent geological period.

Obviously, a climatic change caused by the thawing of polar sea ice will not be limited to an increase in air temperature. A gradual breakup of the Greenland and Antarctic glaciers, leading to an increase in the level of the world ocean, is one possible consequence of the thawing of sea ice. A change in the hydrological cycle on the continents may also be of substantial importance.

We indicated in Chapter 3 that the hydrological cycle significantly changed on the middle-latitude continents of the northern hemisphere when only a comparatively small part of the Arctic ice cap thawed in the course of the warming during the 1920s and 1930s. Changes in the cycle led to an increase in the frequency of drought in a number of regions and to a change in the run-off and levels of land-locked seas.

Many aspects of the climatic change occurring in the first half of the twentieth century were of practical importance. The climatic change that may be reached in the last decade of the twentieth century may therefore strongly influence various aspects of economic activity in a number of countries. A further intensification of climatic changes due to thawing of polar ice may turn into a natural catastrophe if special preventive measures in the national economy are not taken.

These arguments on the climatic condition of the future must be complemented with an estimate of the influence of anthropogenic aerosol on the thermal regime.

Though, as we noted above, it is difficult to predict these changes, it is nevertheless unlikely that the amount of anthropogenic aerosol will significantly grow in the future, since this would lead to conditions highly unfavorable for the life of the population in regions where the most powerful sources of this aerosol exist, that is, in cities and industrial centers. The growth in the mass of anthropogenic aerosol at the end of the twentieth century and, in particular, in the twenty-first century, will thus hardly compensate the tendency toward an increase in air temperature due to a growth in energy production and an increase in the concentration of carbon dioxide in the atmosphere.

If we assume that the amount of anthropogenic aerosol will decrease over the next decades as a result of advances in the control of atmospheric pollution, the process by which polar sea ice recedes and the climatic changes associated with this process will then accelerate.

Anthropogenic changes of the heat and water balances could be accompanied by a change in the amount of oxygen in the atmosphere due to man's economic activity, this oxygen being consumed in the burning of different types of fuel and various production processes (Davitaya, 1971).

Calculations demonstrate that enormous climate changes as a result of the growth in the concentration of carbon dioxide and an increase in energy production, which constitute an additional source of heat for the atmosphere, will occur before the amount of oxygen noticeably decreases. Thus the problem of preserving atmospheric oxygen can arise only at the last stages of an ecological crisis created by man's activity.

These possible changes in the next century are hypothetical in many respects. It is necessary to repeat that the extrapolation used in these calculations of the substantial rates of growth of energy production and carbon dioxide concentration in the atmosphere is only a possible, but not a necessary, condition for economic activity in the future.

Moreover, while further studies may confirm the conclusion that sharp climatic changes are possible in the next century, measures will be taken to limit the influence of industrial development on climatic conditions. This question will be considered in the next sections of this chapter.

We should therefore emphasize that the data presented here were obtained as a result of highly approximate calculations. These calculations are based on highly schematized models of the thermal regime and involve many approximating assumptions. For this reason, the data presented in Figs. 33, 34 and 35 are saddled with significant errors. Though it is rather difficult to estimate the magnitudes of these errors, some conjectures may be made as to the accuracy of the results.

Reliability of Results. The reliability of available estimates of possible changes in the global climate warrants much attention since they could be of practical importance. Two different approaches have been used in treating this question.

The first approach is based on an analysis of the physical foundation of numerical models used for calculating climate changes in the past and in the future, and on a comparison of results of these calculations using different numerical models. The second approach consists in checking the numerical models against empirical data on climatic changes over time.

The semiempirical model utilized here for the thermal conditions of the atmosphere takes into account all the components of the heat balance that influence the horizontal temperature distribution. This model takes into account the most important feedback relations between air temperature and other meteorological factors, including the dependence of air temperature on its absolute humidity (due to the influence of the absolute air humidity on the outgoing radiation, which is allowed for in equation (2.3)) and the dependence of air temperature on the position of polar ice. Both these feedback relations intensify the influence of changes in climate-forming factors on the thermal regime, which decreases climate stability.

Among the assumption made in justifying our model, mention must be made of the loose schematization of the meridional redistribution of heat in the atmosphere and hydrosphere and of the fact that the influence of possible changes in cloud cover on the thermal regime of the atmosphere was not taken into account.

As we noted above, the empirical relation used for a parametric description of meridional heat redistribution holds within a broad range of variation of the distribution of mean latitudinal air temperatures attained in the course of a year. We may thus conclude that the use of this relation will not lead to great errors in calculating changes in the thermal regime smaller than its fluctuations in the course of a year. It is less clear whether this dependence can be used to study drastic changes in the thermal regime due, for example, to the development of planetary glaciation.

The influence of changes in the cloud cover on the thermal regime of the atmosphere has been discussed in many studies, this influence being overstated in a number of cases when no provision was made for a decrease in outgoing long-wave radiation in the presence of a cloud cover.

It is evident from the results of calculations carried out in Chapter 2 that for mean global conditions changes in the cloud cover have a negligible influence on the temperature at the earth's surface, since in this case a decrease in absorbed radiation accompanied by a growth in the cloud cover is nearly compensated for by a decrease in outgoing long-wave radiation. The equations presented in Chapter 2 imply that the influence of a cloud cover on air temperature at the earth's surface will strongly depend on the magnitude of solar radiation. That is, a growth in cloud cover will decrease air temperature for high values of radiation influx to the external boundary of the atmosphere and will increase temperature if the radiation influx is low.

It was established by Schneider (1972) that the influence of cloud cover on air temperature likewise depends on the thickness of the cloud layer, since clouds with a high upper boundary more strongly decrease outgoing radiation. Since the albedo of the earth-atmosphere system increases with increasing thickness of the cloud layer, under particular conditions a decrease in absorbed radiation can compensate a decrease in out-going radiation. Schneider noted (1972, 1973) that the feedback between the temperature field and the cloud cover can influence climate stability, though even the sign of this relation is unknown at the present time.

Some data indicate that changes in cloud cover rather weakly influence the distribution of the mean latitudinal air temperature. This conclusion follows, in particular, from the data of Fig. 36, which depicts the distribution of the mean latitudinal temperature in the northern hemisphere for mean yearly conditions, found by means of the semiempirical model described in Chapter 2 (curve T_0). The influence of deviations of the mean latitudinal values of the cloud cover from its mean planetary value of 0.50 was ignored in this calculation.

Since the mean values of cloudiness at different latitudes are highly variable, the fact that the results of this calculation are quite near the observational data (curve T) indicates the limited influence of changes in the cloud cover on the distribution of the mean latitudinal air temperatures. This conclusion gives us reason to believe that fluctuations in the cloud cover can be disregarded in a number of cases when calculating changes in the thermal regime.

This assumption was further corroborated by studies of correlation between variations in cloudiness and air temperature carried out using satellite data (Budyko, 1975; Cess, 1976).

Let us consider results of applying different models for studying climatic changes.

In Chapter 2 we noted that semiempirical models of the thermal

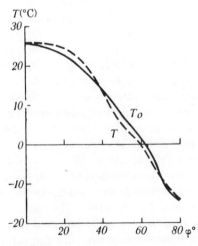

Fig. 36. Distribution of mean latitudinal air temperature. T, observational data; T_0, results of calculations.

regime of the atmosphere that have been developed by various authors yield essentially similar dependence of changes in mean latitudinal air temperatures on external climate-forcing factors. These dependences sometimes reasonably agree both qualitatively and quantitatively, while in other cases there exist quantitative differences between them, which can be explained by the distinctive features in the parameterization of particular atmospheric processes in the corresponding models.

A comparison of results from the semiempirical models and more general models of climate theory that directly postulate circulation processes in the atmosphere is of significant interest.

Earlier chapters compared calculations of changes in the thermal regime of the atmosphere according to the semiempirical model and Manabe-Wetherald's three-dimensional model as a function of fluctuations of the solar constant and of carbon dioxide content in the atmosphere. This comparison demonstrated that results of these calculations are in satisfactory agreement.

Let us consider one more comparison of data from calculations of changes in the thermal regime using more general models and semiempirical models of climate theory.

The change in the air temperature field in the troposphere as a result of the additional heat influx created by man's economic activity has been calculated by Washington (1971) using a three-dimensional model of climate theory. This heat influx, equal to 50 $cal/cm^2 \cdot day$, was uniformly distributed over the surface of the continents.

Washington found that the mean air temperature in this case will grow 1–2° in the tropics and 8–10° in high latitudes of the northern hemisphere.

A calculation using the semiempirical model of the thermal regime of the atmosphere set forth in Chapter 2 yields an increase of 3° in the mean global temperature for these conditions, which agrees reasonably well with Washington's estimates (we note that in both cases the dependence of the albedo of the earth-atmosphere system of temperature changes was not taken into account).

In his next work Washington (1972) calculated temperature changes for an additional heat influx equal to one-sixth of that used in the first calculation, this influx being distributed according to population density. A change in the mean global temperature for these conditions was small compared to its natural fluctuations for periods of several tens of days (these fluctuations, according to Washington's calculations and using the same model of climate theory, amounted to several degrees).

Leaving aside the question of whether such natural fluctuations of the mean global temperature are realistic, let us note that Washington's conclusion agrees with results of calculations using the semiempirical model, which yield a change in mean global temperature of 0.5° for this case, i.e., a magnitude significantly less than the natural fluctuations characterizing his model.

In discussing results of calculations of climatic changes by different models, we should emphasize that these results must be compared in terms of how problems in corresponding studies were formulated.

Sharp differences in several surveys may be thought of as a consequence of particular features of the models used, whereas these differences are basically explained by the fact that the initial values for the calculations in these studies were not equal.

Thus, we may conclude that different models of climate theory yield similar results in most cases when estimating climate fluctuations for identical changes in climate-forming factors.

Some differences in these estimates, occurring, for example, in calculations of the influence of carbon dioxide concentration on air temperature, are apparently often explained not by the particular details of given models of climate theory, but by the features of different atmospheric processes, which are taken into account only parametrically in all climate theories. A second reason for possible divergences in estimates of climate changes consists in differences of the values given in the calculations of such physical parameters as, for example, the albedo of the earth-atmosphere system with and without a snow cover. In a number of cases comparatively small variations of these values lead to significant differences in estimates of changes in the thermal regime.

In verifying the model of the thermal regime used in this book by the data on climatic changes in the past, we find that this model is sufficiently precise (see Chapters 3 and 4).

These results also indicate that this model apparently takes into account all the factors that are sometimes thought to ensure the stability of the contemporary climate. It is easy to understand that otherwise the calculations of contemporary climate changes would give highly overstated values of fluctuations of climatic conditions. However, empirical data demonstrate that the results of calculations are similar to data obtained from observations.

A similar conclusion can be drawn about the major climatic changes that occurred in the Quaternary period.

In discussing the reliability of these estimates of the climatic conditions of the future, we must bear in mind that the laws for climatic changes as man's economic activity rapidly intensifies are of a special nature.

7 THE CLIMATE OF THE FUTURE

In particular, Fig. 34 implies that once the polar ice begins to thaw, the rate of its decrease will itself rapidly increase. This law allows us to obtain a comparatively precise estimate of the time necessary for complete thawing, even assuming great errors in the calculation of the curve shown in Fig. 34. This time will change about 10%, which is insignificant from the practical point of view, if the ordinates of this curve are either doubled or halved. In other words, it is difficult to make a great error in calculating the time at which the polar ice will thaw, even using highly schematic models of climate, due to rapid changes in climatic conditions as the heat barrier is approached.

Various hypotheses concerning the carbon dioxide concentration in the atmosphere in the twenty-first century also scarcely affect the results of the calculation (the hypothesis that the carbon dioxide concentration after the year 2000 will remain constant increases the time within which ice will thaw by a comparatively small value).

Since the results presented here provide only a general picture of the possible climatic conditions of the future, it is necessary to derive new independent data that could be compared with the results presented here.

The need for such data is primarily a function of the great practical importance of future climatic changes. Data on the climate of the future can change the outlook for the development of the most important branches of the national economy in many countries.

These results yield only a general and highly schematized picture of possible climatic changes. Obviously, they must be complemented with more details, including, in particular, estimates of changes of the hydrological cycle in different areas.

The only method of obtaining such detailed data requires extremely general models of climate theory, which have been successfully developed in studies by Smagorinsky, Manabe, and their associates (Smagorinsky, Manabe, and Holloway, 1965; Holloway and Manabe, 1971; Manabe and Wetherald, 1975), in studies by Minz, 1965; Shvets et al., 1970 and other authors.

This problem will probably be solved in the comparatively near future in spite of great difficulties involved in the wide application of models of general climatic theory. The recently derived conclusions that the new global warming has started are of importance for evaluating the reliability of the results of the calculations given here.

In the early 1970s it was universally accepted that over the last decades climate cooling developed. Since the sign of temperature trends changes rather seldom (from the end of the nineteenth century up to the middle of the twentieth century it changed only once), almost all the investigators of climatic variations thought that in the immediate future the temperature lowering would continue. Lamb (1973) mentioned that in the early 1970s there were available more than twenty forecasts of climatic changes which all predicted the cooling to be continued during the next decades. He indicated that those forecasts had no adequate scientific foundations. Only two years after this work had been published Lamb obtained the first data pointing to the fact that the cooling of climate would end (Lamb et al., 1975). These data characterize the thermal regime in the North Atlantic, where beginning with the 1970–71 winter an earlier prevailing tendency for lowering temperature had changed to one of warming.

To study present-day climatic changes, the data on the air temperature of the Northern Hemisphere were supplemented with materials for recent years including 1975 (Borzenkova et al., 1976; Budyko, Vinnikov, 1976). Fig. 37 presents the mean air temperature variation for the Northern Hemisphere for 1965–1975 (the curve A, temperature values are subjected to five-year running averaging except for the last two years). As is seen from Fig. 37, the mid-1960s brought the temperature increase which accelerated in the first half of the 1970s.

The above works present data indicating that warming took place in all latitudinal belts of the Northern Hemisphere, its amplitude increasing with increasing latitude. The warming was accompanied by decreasing mean meridional temperature gradient and lowering the mean precipitation amount in a number of regions of the mid-latitude continents. It was connected with great droughts of 1972–1976 in the vast territories of Europe, Asia and North America. Considerable decrease in the area of polar sea ice in the Atlantic and Arctic was the other consequence of the warming.

The second warming of the twentieth century along with the consequences indicated here was initially predicted (Budyko, 1972b) and only later was it found from observational data. This shows that it is possible to use modern methods of physical climatology for forecasting climatic changes.

It should be noted that Fig. 37 presents data on the most likely mean air temperature variation of the Northern Hemisphere which could occur up to the end of the twentieth century. This assumption is based on computing the CO_2 concentration growth effect upon temperature (Curve T), taking into account the possible influence of the atmospheric aerosol mass variation on temperature which is represented by the interval between the curves T_1 and T_2 (Budyko, Vinnikov, 1976).

It is obvious that this forecast can be checked in the near future from observational data on air temperature.

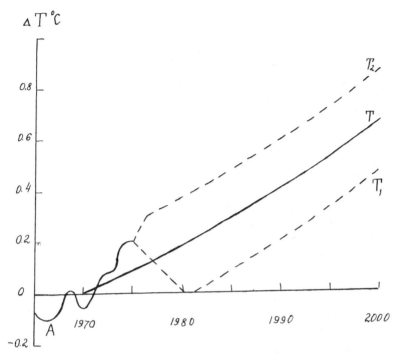

Fig. 37. Secular trend of anomalies of mean air temperature.

7.2 The Possible Ecological Crisis

Ecological Crises of the Past. The balanced nature of the biosphere was the basic feature of the Pre-Holocene epoch. This made possible the prolonged existence of a number of species of living organisms for many millions of years.

Global disturbances of the balanced ecology of the biosphere during critical epochs in geological history that were considered in Chapter 5 were comparatively rare, occurring within intervals of time comparable with the duration of geological periods. Local disturbances of the balanced ecology, resulting from the isolation of ecological system of different scales being violated, were far more frequent. An example of one such disturbance at the end of the Pliocene after North America had been joined to South America by the Isthmus of Panama was mentioned in Chapter 5; an exchange of many members of the fauna of these continents then took place.

Though changes in environmental conditions strongly influenced the evolution of living organisms, great disturbances of the balanced ecology of the biosphere were comparatively rare prior to the appearance of man, which is explained by the high stability of Pre-Holocene ecological systems because of the mutual adaptation of their organisms.

It is well known that man's contemporary activity greatly influences many natural processes forming the geographic medium throughout nearly all regions of the world.

The flora and fauna of our planet have varied quite strongly. Many animal species have been completely exterminated by man, and an even greater number of species exist with sharply decreased populations and are in danger of becoming extinct. Characteristically, these species include not only nearly all the large land animals, but also many species of marine animals, including sea mammals, which have suffered particularly from ruthless pursuit by man.

The plant cover of dry land has experienced enormous changes over the greater part of the continents. Wild vegetation has been destroyed and replaced by farm land over large areas, the forests still existing being mostly secondary growth, i.e., they differ highly from the natural plant cover as a result of man's influence. Great changes have also occurred in the plant cover of many steppe and savanna regions due to intensive grazing by domestic cattle.

Man's influence on the natural plant cover has been particularly acute on the soil development process in specific regions and has led to a change in the physical and chemical properties of the soil. The soil of farm land has been altered to an even greater extent as a result of its systematic cultivation, the use of fertilizers, and the removal of a major portion of the biomass of growing plants.

The influence of man's activity on the hydrological regime of dry land has grown rapidly. The run-off of, not only small, but also many large, rivers has substantially changed as a result of human control through the creation of water engineering works. A large part of the water from river run-off has been removed to provide for the needs of industry and the urban population and to irrigate farmland. The building of man-made lakes, whose area in some cases is comparable to the area of natural seas, has sharply changed evaporation and drainage conditions over great areas.

Man's pollution of the environment, which has spread to the waters of the continents, oceans, and atmosphere, is assuming an ever-increasing scale.

Until recently, the viewpoint prevailed that man's effect on the environment began to manifest itself at greater intensities basically over the most recent decades when industrial development sharply increased and the size of the population of our planet began to grow rapidly.

Recent studies have demonstrated that even thousands of years ago man's activity led to great changes in the natural environment, which sometimes threatened the continuing existence of human society. The first such case apparently resulted from the development of Upper Paleolithic culture in Europe, Asia, and America.

It is known that the Upper Paleolithic was the first culture created by modern man (Homo sapiens sapiens). The economic basis of this culture lay in hunting large animals by using tools that enabled man to capture even such animals as the mammoth and woolly rhinocerous.

The Upper Paleolithic culture lasted quite long—more than half the entire period of existence of modern man equal to 35,000–40,000 years.

As we noted in Chapter 5, many large animals that had inhabited Europe, Asia, and America became extinct during the Upper Paleolithic, which coincided in time with the last Würm. A number of researchers have suggested that these animals were exterminated by Upper Paleolithic hunters.

We should, however, note that most authors doubt that human activity could have exerted a substantial influence at such an early epoch on the disappearance of such large and widespread animals.

It has been noted in a survey of the literature on this question (Butzer, 1964), that the animals that became extinct were highly specialized and thus not able to adapt to changes in the natural conditions during the late Würm. Butzer also noted that a number of predators that had not been the object of systematic hunting by primitive man also became extinct at the end of the Pleistocene, at the same time as did the large herbivores.

In addition, Butzer asserted that not all the hoofed animals of Europe disappeared at this time and that until recently many herds of large herbivores remained in tropical lands. They continued to live and multiply in spite of being intensively hunted by the local population. These considerations, in Butzer's opinion, contradict the viewpoint that man could have exterminated a number of animal species at the end of the Upper Paleolithic.

Nevertheless, Butzer recognized that the abundant remains of large animals in many campsites of the Upper Paleolithic indicate that man could have exerted a definite influence on the total balance of biomass of now extinct animals.

Colbert (1958, p. 459) has written, in discussing the causes for the extinction of mammoths, "Why did the mammoths become extinct? This question, like so many of the questions having to do with problems of extinction, is extremely difficult to answer. In fact it is probable that we shall never know the real reason for the disappearance of the mammoths a few thousand years ago, after their successful reign through Pleistocene time. Very likely the extinction of the mammoths was the result of complex causes. Man may have had something to do with it, but we can hardly believe that primitive man was of prime importance in bringing an end to these numerous, widely spread, and gigantic mammals."

It should be noted that existing studies have limited their analysis of reasons for the marked change in the animal world at the end of the Upper Paleolithic to qualitative statements difficult to prove or refute without a quantitative treatment of the problem. For example, the high specialization of animals that became extinct at the end of the Upper Paleolithic nevertheless did not interfere with their existence under the sharp changes in the environment that repeatedly occurred throughout the Pleistocene.

The extinction of large predators at the same time as that of a large number of herbivores does not prove the absence of any influence of man's activity on the change in the animal world, since the disappearance of these predators can be explained by a disturbance in their food chains as specific herbivore became extinct both as a result of natural causes and as a consequence of man's activity.

The herds of large herbivores that still remain in the tropics do not convincingly demonstrate that similar animals could have been exterminated in Europe by Upper Paleolithic hunters.

Let us discuss the possible influence of man's activity on the extinction of animals at the end of the Pleistocene on a quantitative basis, taking into account the basic factors influencing animal populations (Budyko, 1967).

If such a calculation confirms that this activity substantially influenced the process of extinction, other factors, including climatic changes, may be considered secondary.

Modern man first appeared in Europe about 35,000 to 40,000 years ago, where he replaced the Neanderthal man and created an effective hunting system for large herbivores based on a more advanced technique of processing stone and bone tools. The population of Europe greatly increased during the Upper Paleolithic, the level of material culture noticeably increased, and the first steps were made toward the development of imitative arts (for example, the famous paintings of animals in the caves of France and other European countries). The Upper Paleolithic ended 10,000 to 13,000 years ago, which approximately coincides with the end of the Würm glaciation.

The end of the Upper Paleolithic corresponded to a significant change in the pattern of life of the European population, indicated by the spread of the Mesolithic culture of the Middle Stone Age; many outstanding achievements of Upper Paleolithic culture, including the art of cliff paintings, were lost to some degree at this time with the appearance of a new technique of making stone tools.

The end of the Paleolithic may be said to be characterized by a sharp break in the early history of mankind. The break was apparently due to the cessation of systematic hunting for large herbivores, as a result of which Upper Paleolithic man could no longer provide himself with food and many important materials. Consequently, it was necessary for many to find new sources of supply for his existence.

Let us present here a computational scheme, allowing us to estimate how hunting influenced the population of pursued animals during the Upper Paleolithic.

The scheme is based on an equation that characterizes the variation in the population of an animal species over time:

$$\frac{dn}{dt} = \alpha n - \beta n, \tag{7.1}$$

where n is the population of animals per unit area, dn/dt is the rate at which population changes over time, α is the birth rate (relative increase in population as a result of the birth of animals), and β is the death rate (relative decrease in population as a result of their death per unit time).

The parameters α and β may depend on the population of animals per unit area, i.e., on n, and these dependences may differ for different animal species.

Calculations based on equation (7.1) as well as on field studies have demonstrated that the population of different animal species usually fluctuates under natural conditions within rather broad ranges. These fluctuations have periods commensurate with the lifetime of a single animal generation, which can be explained by changes in food conditions caused by the weather conditions, outbursts of epidemics and complex interactions between the populations of herbivores and predators, which have been shown by Volterra to be of the auto-oscillating nature in a number of cases (Severtsev, 1941; Laek, 1954).

In addition, the mean population of a given animal species will be more or less constant over a period of time markedly greater than the mean lifetime of the animal under constant (on the average) external conditions. Changes in this mean population can occur either as a result of a change in the natural conditions or in conjunction with the evolutionary development of the given animal species or the species interacting with it (Severtsev, 1941; Shmal'gauzen, 1946).

The laws of natural variations of an animal population considerably change when the animals are being systematically hunted. In this case equation (7.1) can be represented in the form

$$\frac{dn}{dt} = \alpha n - \beta_1 n - g, \qquad (7.2)$$

where g is the number of animals killed by hunters within a given period of time.

Obviously, the coefficient β_1 that characterizes death from natural causes will be less than the coefficient β, other conditions being equal, for animals that were pursued by hunters during the Paleolithic, since hunters could more easily capture sick and weakened animals than healthy and strong ones.

Since the products of hunting were the chief food source of Paleolithic man in Europe, we should expect that $g = \gamma m$, where m is the population of man inhabiting a unit area, and γ is the relative amount of biomass of a pursued animal consumed by a single person within a given period of time (ratio of weight of consumed biomass to mean weight of a single animal).

Thus equation (7.2) becomes

$$\frac{dn}{dt} = \alpha n - \beta_1 n - \gamma m. \qquad (7.3)$$

We should note that the "principle of interactions" proposed by Volterra for studying the interrelation between populations of predators and herbivores cannot be used to calculate the decrease in the number of animals resulting from systematic hunting.

In accordance with this principle, the number of slain animals can be considered proportional to their population, which is a natural hypothesis for small animals who are able to conceal themselves from a pursuer. Man, after mastering the technique of systematic hunting, was probably always able to find his prey within the area he inhabited, particularly such large animals as the mammoth, who lived in the open spaces of tundra and steppe. Thus, the term m in equation (7.3) can be considered independent of n as long as the number of pursued animals insured the current needs of a particular tribe over its area.

In calculating the population dynamics of large herbivores during the Upper Paleolithic, it is necessary to estimate the variation of population over this period. Obviously, this variation is determined by an equation similar to (7.1), i.e.,

$$\frac{dm}{dt} = am - bm, \qquad (7.4)$$

where a and b are the corresponding birth and death rates.

These coefficients depend on a number of factors, though in view of the lack of suitable data, we will limit ourselves to a rough estimate of the mean difference of these coefficients, $c = a - b$, assuming it to be independent of time and population density.

In this case, we obtain from equation (7.4):

$$m = m_0 e^{ct}, \qquad (7.5)$$

where m is population size at time t, and m_0 is population size at the initial time.

Some idea of the population density in Europe at the beginning and end of the Paleolithic may be gained from data of recent ethnographic studies of tribes at various levels of historical development.

Various authors who have used this method have been able to conclude that the population density in Europe at the end of the Upper Paleolithic fluctuated from roughly 5 to 50 people per 100 km^2. These values must be compared to the much lower density of population of those people that did not master the technique of systematic hunting, which has been assumed to be about one person per 100 km^2 on the average (cf. Braidwood and Reed, 1957; Butzer, 1964).

Since a corresponding increase in population occurred throughout the Upper Paleolithic, which lasted roughly 25,000 years, we find from equation (7.5) that c is between $0.64 \cdot 10^{-4}$ and $1.56 \cdot 10^{-4} yr^{-1}$.

These coefficients for the population increase, extremely low from the modern point of view, are characteristic for early periods of human history. Their values agree reasonably well with the low mean lifetime of Upper Paleolithic man established from studies of their skeletons.

Let us now calculate changes in the population of a species of large herbivores that inhabited the European tundra during the Upper Paleolithic.

If this species dominated other herbivore species in a given locality, it could use the basic mass of food resources, which, according to available data, was sufficient for maintaining an animal population in the tundra corresponding to roughly 800 kg of biomass per 1 km^2.

This value determined the upper limit for the possible number of these animals. We may assume that this limit was reached under favorable conditions for such animals as the mammoth, whose population was nearly independent of the activity of predators.

Let us consider how the number of mammoths must have varied once man had begun hunting for them on a broad scale. This calculation requires that we solve an equation obtained from (7.3) and (7.5):

$$\frac{dn}{dt} = \alpha n - \beta_1 n - \gamma m_0 e^{ct}. \tag{7.6}$$

Multiplying all terms of the equation by the mean weight of a single animal, we arrive at the form

$$\frac{dN}{dt} = \alpha N - \beta_1 N - \Gamma m_0 e^{ct}, \tag{7.7}$$

where N is the biomass of the given animal per unit of occupied area, dN/dt is the rate at which this biomass varies over time, and Γ is the biomass of pursued animals consumed by a single person per year.

This equation must be solved under the initial condition $N = N_0$ at $t = 0$, under which the biomass of the animals at the start of the Upper Paleolithic was equal to the limit determined by feeding conditions.

In order to simplify the calculation, we will assume that the right side of equation (7.7) vanishes when $\Gamma m_0 e^{ct} < \alpha N_0$, since it is difficult to estimate the coefficient β_1, i.e., the mean number of animals remains constant if losses from hunting are less than the increase in biomass. Then we shall neglect the term $\beta_1 N$ when $\Gamma m_0 e^{ct}$ increases to values greater than or equal to αN_0, and we will use equation (7.7) in the form

$$\frac{dN}{dt} = \alpha N - \Gamma m_0 e^{ct}. \tag{7.8}$$

Obviously, both these simplifications lead to some understatement in the calculation of the rate at which an animal population decreases in comparison with the actual state of affairs.

We can determine this value from the equation

$$\Gamma m_0 e^{c\tau} = \alpha N_0, \tag{7.9}$$

denoting by τ the time during which hunters scarcely affected the number of animals they pursued, from which we find that

$$\tau = \frac{1}{c} \ln \frac{\alpha N_0}{\Gamma m_0}. \tag{7.10}$$

We must solve equation (7.8) in order to determine the length of time during which the mammoth population decreased and finally became completely extinct.

7 THE CLIMATE OF THE FUTURE

Assuming $N = N_0$ when $t = \tau$, we obtain a dependence of the biomass reserve of given animals on time:

$$N = \frac{\Gamma m_0}{\alpha - c} e^{ct} + e^{c(t-\tau)} \left[N_0 - \frac{\Gamma m_0}{\alpha - c} e^{c\tau} \right]. \tag{7.11}$$

From this equation, we find the period of time after which N is equal to zero, which corresponds to the extinction of the given animal species.

A calculation using these equations of changes in the mammoth population during the Upper Paleolithic requires, in addition to the parameters whose values are given above, knowledge of the values of the coefficients α and Γ. The first coefficient characterizes the ratio of number of mammoths born in a given year to the total population of the herd. This coefficient is 0.05 according to data for modern elephants.

The coefficient Γ can be approximated from data of ethnographic observations for northern tribes that chiefly feed on large animals. Since Γ characterizes the direct use of biomass for food and for economic purposes as well as unproductive losses, it may be estimated as roughly 500–1000 kg/yr per person. This figure for biomass consumption has been considered somewhat overstated by Bibikov (1969), who accepted the basic conclusions of our calculation. Though we agree that the actual use of animal meat and fat was much less than this value, our estimate is nevertheless highly probable due to significant biomass consumption for economic purposes (animal fat, hide, and bones) and due to its great losses, particularly associated with the difficulties of prolonged storage.

We calculated the variations of mammoth population for two values of c (i.e., $0.64 \cdot 10^{-4}$ and $1.56 \cdot 10^{-4}$ yr^{-1}) using the above equations and assuming that $\Gamma = 750$ kg/yr and letting $N_0 = 800$ kg/km^2 and $m_0 = 0.01$ km^{-2} in accordance with the above estimates.

If we calculate τ using equation (7.10), it becomes apparent that the mammoth population varied comparatively little for 10,000 to 25,000 years (depending on the given value of c) after they began to be hunted. The mammoth population then began to decrease rapidly as is evident from equation (7.11), and the mammoths became completely extinct within several centuries after the end of the period of stable population.

In evaluating the results of this calculation, we must bear in mind that its precision is limited by the approximate nature of the parameters used. In addition, our equations can be seen to imply that significant variations in the values of the parameters do not change the general laws of the process and do not weaken the basic conclusion that such large animals as the mammoth could have become extinct within a period of time commensurate with the duration of the Upper Paleolithic.

It can be easily demonstrated that this conclusion does not change if we replace the dependence of population on time used above by simpler linear relations. We note that our conclusion is confirmed also whenever a given animal species (for example, the mammoth) is not the only herbivore subsisting on a given food base. For this purpose we must assume that the ratio of the biomass of a given species to the total biomass of the pursued animals approximately corresponds to the similar ratio for the biomass of animals slain by hunters.

Our conclusion can also be obtained without any complicated calculations, simply by comparing available data on the biomass balance of large herbivores in Europe during the Upper Paleolithic. A comparison of the annual increment of this biomass to the size of the population occupying this area is of great importance here.

If the increment was at most 4000 kg per 100 km^2, as in the case of the mammoth, it becomes obvious that it was sufficient to maintain a population of at most several people per 100 km^2 even if it was completely used by the hunting tribes. Clearly, however, the population will have continued to grow after reaching this size, since for a long time there was a sufficient amount of food not only due to the increment in biomass, but also as a consequence of the extermination of part of the main herd of hunted animals. Finally, prior to complete extermination of the animals, the increment in biomass will gradually decrease, which is in accordance with the results of our calculation.

We may suppose that once the mammoths had been eliminated by man over most of man's area, subsequent herds could have been left in underpopulated regions in Northern Asia. However, in regions with such a severe climate, large herbivores could have a supply of food only during relatively warm periods. A succeeding climatic change toward cooling would have led to the extinction of the mammoths in these regions as well.

Using these equations, we may explain why primitive man was able to eliminate the mammoth but not the kindred elephants, which until recently were numerous in a number of tropical regions.

It is evident from equation (7.10) that the time an animal species has left to it, if it is being hunted down, depends on its possible biomass reserve Γ and the relative use of this biomass by hunting tribes. Available data show that the biomass of large herbivores in tropical savannas is 10 times the corresponding biomass in tundra. In addition, in tropical latitudes primitive man's food could have included a much larger amount of vegetation than did the food on which Paleolithic hunters in Europe subsisted. Here it was limited to game, due to glaciation. Since, for hunters in the tropics, Γ is at least half the value we have assumed for Upper Paleolithic conditions in Europe, and taking the value of N_0 corresponding to savanna conditions, we find from equation (7.10) that for the tropics, τ must have been roughly three times that found above for Europe during the Paleolithic epoch.

Thus the population of elephants, even when constantly hunted, could have been maintained at a high level for several tens of thousands of years after mammoths had been exterminated.

This theory allows us to explain why only the largest herbivores became extinct in Europe during the Upper Paleolithic. Since the population of any species of these animals was limited by the food resources of the given geographic zone, the increment in biomass of a species was thereby determined in turn by the birth rate α. It is well known that birth rate is minimal for the largest animals and increases with decreasing animal size. Species of large herbivores whose biomass increment per unit area was least were therefore the least protected against Upper Paleolithic hunters. The causes of the biological catastrophe that overtook this species is therefore quite simple.

The increase in the size of large herbivores in the course of their evolution freed them to a certain degree from the threat of predatory attack, which made a significant decrease in birth rate both possible and expedient. When Upper Paleolithic man began to hunt for these animals, he put himself in the role that wolves and other predators hold toward the smaller herbivores. However, while the smaller herbivores could compensate for their losses by means of a high birth rate, this was not possible for large animals, and it was a matter of time before their extinction by pursuit, a pursuit for which their evolution had not adapted them.

It should be noted that this explanation differs from the one given by Simpson (1944), who suggested that large animals experienced greater difficulty in adapting to the changing natural conditions of that time due to their longer lifetime and a comparatively slower change of generations.

The actual period of time from the start of systematic hunting to the point at which large herbivores began to die out in different regions of Europe was probably significantly briefer than that which can be obtained from a calculation using our scheme. This discrepancy is due to the assumptions made in the calculation and also to the fact that fluctuations in animal population as result of a epizootic diseases and, in particular, changes in the food base, which could have been extremely significant under the conditions of the last Würm, were not taken into account in the calculation.

The existence of such fluctuations could have accelerated the disappearance of various large herbivores within a fixed region over the course of many thousands of years. On the other hand, it would have been difficult for a species to re-establish its population in this region by migration from other regions, since Upper Paleolithic hunters would have continued to pursue it.

From this point of view, the largest animals with lowest birth rate would have vanished before other animals in the course of systematic hunting. Their disappearance would have facilitated a temporary increase in the population of smaller herbivores, their food base increasing. However, a further increase in the rate of hunting could have led to the disappearance of herbivores with a high birth rate as well, which was, in fact, the case of reindeer in a number of regions of Western Europe. We should mention here that losses due to man's pursuit for these animals were complemented by substantial losses due to pursuit by predators, which reduced the stabilizing effect of a high birth rate for maintaining the population of a species.

A number of other results of this computational scheme should also be noted.

The process in which the population of pursued animals decreased was at first slow and gradually accelerated, and the last, still comparatively numerous herds of animals were exterminated quite rapidly. In addition, more or less long periods during which pursued animals were absent as a result of natural fluctuations of their population, migration of different herds, and so on must have occurred long before complete extermination of any animals.

Thus, as the population of Europe increased during the Paleolithic, the existence of this population, which basically depended on systematic hunting, was quite often threatened. This accounts for the unusual attention Upper Paleolithic man paid to animal paintings, which were of the nature of magic, as a way of reviving vanished herds. However, the general process of extermination of large herbivores increasingly accelerated with systematic hunting and its final phase was particularly rapid. This put the hunters of a tribe in a difficult position, since they lacked sufficient time to gradually switch over to other methods of food gathering. Thus the transition to the Mesolithic was probably a painful experience for primitive society and may have been accompanied by a temporary decrease in the population.

As we noted above, the end of the Paleolithic in America was also accompanied by the extinction of large animals, which confirms the causal relation between these phenomena. It is also entirely likely that important changes in natural conditions toward the end of the Paleolithic presented more difficulties for the existence of animals that were becoming extinct and helped to accelerate an already ongoing process.

Thus Paleolithic culture ended in Europe partly as a result of an insoluble conflict between the technique of systematic hunting of large animals that Upper Paleolithic man had invented and which both ensured a temporary abundance of food and made possible an increase in population, and the limited natural resources that could be hunted in this way and which were in fact exhausted in the course of time.

We may therefore ask ourselves what could have disturbed the balanced ecology of the Pre-Holocene biosphere, an ecology in which animal species pursued by predators may have existed for hundreds of thousands and millions of years.

The chief cause of this disturbance may have been man's unusually high rate of evolution in comparison with the rate of evolution of his prey.

Large animals with a low birth rate were less able to maintain their population at a stable level when pursued by predators than were small animals with a higher birth rate.

Large body size therefore gave animals an advantage in the struggle for existence only when it made animals less accessible to attack by predators. But if animal size did not protect them from attack (as was the case even for mammoths during the Upper Paleolithic), an increase in birth rate would be the only method for maintaining their population.

Such a change of pursued animals at the rate of evolution characteristic of the Pleistocene required far more time than the period it took for their extermination.

We noted in Chapter 5 that a similar approach was used by Martin (1973) to study the extinction of many species of large animals in North and South America at the end of the Pleistocene. It is known that man appeared on these continents comparatively late (at the end of the last Würm glaciation), and that the process by which America was settled has not been studied very thoroughly. Martin proposed that a small group (about 1000 people) crossed the ice of the Bering Strait about 11,000 years ago and reached the northwestern regions of North America, where they found enormous herds of large animals that had never encountered man and which lacked protective instincts. Thus the hunting conditions were more favorable in America than in Europe or Asia, where wild animals had been in contact with man at various stages of his evolution for hundreds of thousands of years, during which time these animals had developed various protective mechanisms against pursuit by hunters.

As a result, the first small group of inhabitants of America had an unlimited amount of food at their disposal and, after rapidly increasing their population, settled all of America, exterminating most of the large animals they encountered. Martin's calculations demonstrated that the wave of first settlers could have reached Tierra del Fuego 1000 years after completely exterminating about 30 genera of large animals still existing in America at the end of the Pleistocene (mammoth, mastodon, megatherium, horse, camel, and many others). The hunters of the tribes were forced to switch over to other methods of food gathering once these animals had been exterminated, which led to a sharp decrease in the population of America.

This problem was subsequently studied by Mosiman (personal communication), who investigated the time at which large animals pursued by Upper Paleolithic hunters became extinct, as a function of variations of the parameters and the relations of this model. He thereby established that the basic conclusions obtained in these calculations remain in force whether the parameters of the model are varied within broad limits or if a number of modifications are introduced into the model itself in order to refine its initial assumption.

Results of these investigations have indicated what was the probability that an ecological crisis materialized at the end of the Upper Paleolithic, spreading over a large portion of the area populated by man. This crisis apparently led to a decrease in man's population wherever the large animals that had been the basis for the Upper Paleolithic economy became extinct.

It is quite possible that in a number of cases this crisis resulted in a gap between the end of Upper Paleolithic culture and the emergence of Mesolithic culture, which has been noted by archaeological investigations in Europe and other regions.

The idea that the first and longest culture in the history of modern man ended in an ecological crisis is of interest as an illustration of the theoretical incompatibility between man's spontaneous activity and the maintenance of a balanced ecology. Other cases are well known when the balanced ecology of the biosphere was disturbed in the most recent epoch in the history of the development of human society. These cases confirm the statement made by K. Marx that culture, if it has developed spontaneously and is not consciously directed, leaves a desert after itself (K. Marx and F. Engels. Letters. Works, vol. 32, Moscow, 1964, p. 45)

Ecological Crisis of the Future. Though man's contemporary activity, like his activity in the past, has substantially changed the environment over a large part of the planet, until recently these changes were only the sum of many local affects on natural processes. They assumed a planetary nature not as a result of man's alteration of natural processes on a global scale, but because local (or regional) effects have spread over large areas. In other words, a change of the fauna in Europe and Asia did not affect (or affected only slightly) the fauna of America, control of the run-off of American rivers did not change the run-off of African rivers, etc. Man's effect on global natural processes began only very recently and this change has influenced the natural conditions of the entire planet (Gerasimov and Budyko, 1974).

The hypothesis has been recently put forward that a further development of man's economic activity could markedly change the environment, leading to a general crisis in economics and a sharp decrease in man's population.

This point of view has been substantiated in greatest detail by Forrester (1971) and Meadows et al. (1972).

As was done in a previous work dealing with the economics of the Upper Paleolithic (Budyko, 1967), Forrester and Meadows and his colleagues have solved an equation that determines the increase of population consistent with a change in the resources utilized by man. Like the calculation for the Upper Paleolithic, the analysis carried out by the American scientists for the modern era led to the conclusion that a further growth in population will issue in a crisis that will sharply decrease population. According to this study, such a crisis will manifest itself in only about 100 years.

Models for the modern era have taken into account a number of factors influencing population size, including food and mineral resources and environmental pollution. Since the model for the Upper Paleolithic was limited to food resources (reserves of mineral raw materials did not then affect population and environmental pollution did not exist at all), a calculation using this model could be carried out analytically. The far more complex systems of equations used by the American scientists could be solved only numerically by using computers.

Let us discuss in detail the results of the prediction of changes in economic conditions based on the use of Forrester's model.

This calculation was carried out for the period between 1900 and 2100 and used dependences of birth and death rates on various factors obtained from contemporary empirical data.

When only scattered data were available (for example, for estimating the influence of environmental pollution on the death rate), the corresponding dependences for the calculation were hypothesized.

Results of the calculation demonstrated that industrial production and the production of food per capita will grow through the beginning of the twenty-first century, after which they will rapidly fall far below their level at the beginning of the twentieth century.

Reserves of mineral resources at the beginning of the twenty-first century will decrease sharply and will be practically exhausted by the middle of the century. Environmental pollution will rapidly grow in the first half of the twenty-first century and will again decrease toward the middle of the century due to a drop in industrial production.

Population will grow through the middle of the twenty-first century. It will then begin to decrease, and will be nearly halved by the end of that century. The exhaustion of mineral resources, which will lead to a sharp increase in the death rate due to a decrease in the food production and a worsening in the medical care of the population, will be the principal factor in this crisis.

Results of this calculation in the book *Limits to Growth* were compared to a number of other calculations that employed various additional assumptions. However, in most cases these assumptions did not change the basic conclusion obtained in the initial calculation.

For example, the assumption that currently known reserves of mineral resources will double leads to a still sharper drop in population by the middle of the twenty-first century, since in this case the level of environmental pollution will catastrophically increase.

The possibility that environmental pollution will be effectively controlled does not exclude a drop in population due to insufficient food resources.

Results from subsequent calculations show that a complex of measures is required in order to keep population from decreasing and to maintain a high level of industrial production. These include a halt in the growth of population, a sharp decrease in the use of mineral resources, a cessation in the growth of industrial production, strict control of environmental pollution, an increase in food production, and others.

Such calculations imply that measures for stabilizing economic conditions in the twenty-first century must be introduced rapidly, and not put off to the end of the twentieth century when it will be too late to prevent the onset of the crisis.

These authors have indicated that the impending economic crisis may end in "social disintegration," i.e., in the breakdown of the existing organization of society, since it does not seem very realistic to assume that these measures will be carried out in the modern era.

A critical evaluation of these studies has been given by Fedorov (1972). In his book he noted the fundamental drawbacks of the above studies, including especially the fact that the numerical models used do

7 THE CLIMATE OF THE NUTURE

not take into consideration those qualitative changes that are introduced by technical progress into the economic development of society and into the manner in which man's demands are satisfied.

In addition, Fedorov indicated that the conclusions of the American authors regarding the impending crisis in the interaction between society and nature correctly reflect the laws of spontaneous development of bourgeois society. The fundamental social and economic features of the capitalistic system prevent the rational use of natural resources, hamper the solution of the food problem, and complicate the systematic struggle against environmental pollution. Under such conditions, a crisis in man's interrelation with nature is unavoidable and many features of this crisis can be clearly seen in the modern era.

Fedorov notes that the difficulties that will arise with mankind's technological and economic development can be overcome only through a planned socialist economy.

The conclusion that long-term economic planning is necessary can also be found in the work of American authors, though the measures proposed here for preventing the onset of the crisis are clearly Utopian.

It may be possible to find more realistic ways of overcoming these difficulties even within the framework of the limited numerical models used in their studies. For example, simple calculations demonstrate that most of the obstacles to technological and economic growth will have been eliminated in the next century by greater control of environmental pollution, massive use of synthetic materials, and at least partial advances in the chemical synthesis of food products.

It is necessary to study how currently available methods of numerical modeling of economic development can be perfected in order to solve particular problems in long-term planning of technical progress.

We can formulate in advance these mechanical problems by using numerical models that allow us to predict probable changes in population as economic development proceeds and as natural conditions change.

One of the most important of such problems is that the global climate of our planet may be modified as a result of economic activity, a question that was not considered in the American studies.

This is particularly important, since such a change will substantially influence man's economic activity earlier than any other global ecological disturbances.

This influence will be felt on a large scale. We need only note that the warm period of the 1920s and 1930s was accompanied, as can be seen from the data of Chapter 3, by a noticeable decrease in the amount of precipitation in a number of intracontinental regions. This substantially worsened conditions for agricultural production, river transport, and other branches of the national economies of a number of countries.

The data of this chapter imply that the influence of man's economic activity on climate can lead in the comparatively near future to a warming comparable with that of the first half of the twentieth century, and will then greatly exceed it. Thus, a climatic change can be the first indicator of the global ecological crisis mankind will encounter in the course of fundamental developments in technology and economics.

The redistribution of the amount of precipitation in various regions of the world, in conjunction with a noticeable decrease in many regions of unstable humidity, will be the chief cause of this crisis in its first stage. Since most grain cultures are grown in these regions, a change in precipitation conditions can substantially complicate the problem of providing sufficient food supplies for the rapidly increasing global population.

For this reason, the task of overcoming undesirable changes in global climate becomes one of the most important ecological problems of the contemporary age.

7.3 Climate Modification

Methods of Affecting Climate. Different measures have been taken in order to minimize unfavorable climatic changes arising from man's economic activity; atmospheric pollution is now being controlled on a broad scale.

It should be noted that the level of air pollution has recently decreased in a number of cities as a result of numerous measures taken in many developed countries, including air purification at industrial plants, on vehicles, by heating units, and so on. However, in many regions air pollution has increased. This indicates how great are the difficulties involved in preventing a growth in the amount of anthropogenic aerosol in the atmosphere.

It would be even more difficult (and the relevant problems have yet to be raised) to prevent an increase in the carbon dioxide content in the atmosphere and a growth in the heat released in transformations of the energy used by man.

There are no simple technical means for solving the first and, especially, the second problem, other than limits on the use of fuel and most types of energy, which is probably incompatible with further technical progress over the next decades.

Thus in the near future climate modification will become necessary in order to maintain current climatic conditions. Obviously, if this becomes possible, it could also prevent natural climatic fluctuations unfavorable to the national economy, and, subsequently, modify the global climate in the interests of mankind.

Possible ways of affecting climate have been discussed in a number of works containing different ideas for modifying climatic conditions.

It has been repeatedly noted by Fedorov (1958, 1972) that modifications of large-scale atmospheric processes characterized by low stability, i.e., that are highly sensitive to comparatively small changes in external factors, are the most promising. Such an approach has yielded good results in developing methods for dealing with clouds, precipitation, fog, and other weather phenomena.

Fedorov gave an interesting example in which the radiation balance of a large area was noticeably changed by cloud seeding. Similar experiments have demonstrated that weather modification can change climatic conditions if applied over a long period of time. Thus Fedorov stated that it was theoretically possible to modify the climate of the entire globe or a significant part of it.

Various projects of weather modification are exploited in a number of studies.

One of the largest such projects was intended to eliminate Arctic ice in order to increase the air temperature at high latitudes. A number of studies of the relation between the polar ice regime and general climatic conditions have been carried out to deal with this question (Budyko, 1961, 1962; Fletcher, 1966).

The results of these studies do not exclude ice-free conditions once the polar ice is eliminated, but apparently indicate that such conditions are highly unstable.

It should be noted that very few methods of actually eliminating polar ice have been developed. A number of calculations have shown that Arctic sea ice may thaw within several years if the cloud cover decreases in the northern part of the Atlantic ocean and if the rate of evaporation from the ocean surface decreases here. Obviously, though there are certain methods of controlling clouds and evaporation from water reservoirs, it is quite unrealistic to expect that these actions will be carried out over vast areas. Some authors believe that polar ice can be thawed by building a dam in the Bering Strait together with a powerful system to pump water from the Bering Sea into the Arctic Ocean. Other authors suggest that polar ice can be broken up using atomic energy.

As we indicated above, the disappearance of polar ice will have a complex influence on climate and will not be favorable in all respects to human activity. Not all the consequences of this process can now be predicted to a sufficient degree of accuracy. We may therefore suggest that even if it becomes possible to thaw the polar ice cover, it will not be advisable to do so in the near future.

Methods of altering large-scale atmospheric movements is another possibly fruitful way of modifying climatic conditions. Until recently, it was assumed that movements of air masses involve such great energy consumption that man's influence on them could not yield any results at the present time. However, in many cases atmospheric motions are unstable so that they can be affected using a comparatively small amount of energy. It is thus of interest to determine how the intensity of a low-level cyclone could be changed by using a system of vertical air flows created over the cyclone-forming area. These calculations imply that such effects are theoretically possible at the contemporary level of development of energetics, though the technical means for it can be described only in most general terms (Yudin, 1966).

An analysis of a number of possibilities for climate modification was generalized in the summary report of the Panel on Weather and Climate Modification of the Committee on Atmospheric Sciences of the USA (Weather and Climate Modification, 1966). Some methods of microclimate modification referred to in this report related to agrometeorological problems. These include different methods of protecting plants from night frosts, shading plants in order to protect them from overheating and excess moisture evaporation, planting of forest strips, and so on. Obviously, these methods will not affect the climate of large areas, and, except for protective forest strips, they do not ensure a permanent change of local climate.

The idea of modifying the climate of dry regions was considered for the case of the deserts of southern Peru, which are situated near gently sloping mountains. Air descending along these slopes is heated and becomes less saturated, which impedes condensation and precipitation. It was proposed that the desert surface be irrigated with a small amount of water in the morning, which would lead to a drop in surface temperature. This would result in an improvement of conditions for the development of daytime convective cloudiness and precipitation.

Another plan for modifying precipitation conditions was suggested for deserts with air characterized by a high dust content. The air is cooled because of the effect of masses of dust on long-wave radiation, which impedes the development of ascending currents and decreases precipitation. It was suggested that the dust content of the atmosphere could be decreased in these deserts by restoring the plant cover on their surface, after which a long-term re-establishment of the plant cover will be made possible by an increase in the amount of precipitation. Such a project was developed for the deserts of northwestern India, which are apparently the result of the destruction of the plant cover by excessive cattle grazing (Bryson, 1972).

Other plans for climate modification have been mentioned in a number of publications. These include affecting ocean currents through the construction of enormous dams. For example, it has been proposed that the regime of the Gulf Stream could be modified by constructing a dam between Florida and Cuba, or that the direction of the Labrador Current could be modified by constructing a dam on the Newfoundland Bank. None of these plans have been sufficiently argumented from a scientific point of view and it remains quite unclear how they may influence climate.

Other plans include proposals for large water-storage basins. Among them is the old idea of creating enormous seas in Central Africa and the Sahara by damming up a number of rivers in Africa. These seas would occupy 10% of the total area of the African continent. We should note, leaving aside the question of whether such a project is possible, that none of the climatic changes that could result have been studied to any great extent.

It is probable that some of these plans of climate modification over limited areas will become feasible for the technology of the near future if they prove to be profitable for mankind.

Far greater difficulties are encountered in modifying the global climate, i.e., the climate of the entire planet or of a large part of it.

A method based on increasing the aerosol content in the lower stratosphere is apparently the most accessible one for modern technology (Budyko, 1972b, 1974; Budyko et al., 1974).

Modification of the Aerosol Layer of the Stratosphere. It has been established in a number of studies by Junge (1963), Rozenberg (1968), Cadle (1972), and Karol' (1973) that the lower stratosphere contains an aerosol layer whose particles measure tenths of a micrometer and which consists chiefly of sulphur compounds.

Estimates were given in Chapter 3 for the influence of this layer on the flux of short-wave radiation that reaches the earth's surface. These estimates imply that comparatively small fluctuations of the aerosol mass in the lower stratosphere noticeably vary the meteorological solar constant, and, correspondingly, the mean temperature at the earth's surface.

Let us consider the possibility of changing climatic conditions by increasing the concentration of aerosol particles in the lower stratosphere. For this purpose we will calculate the aerosol mass which has to be ejected into the stratosphere in order to induce a climatic change of the scale observed during the warming period of the 1920s and 1930s, but in the opposite direction, i.e., ensuring a drop in mean temperature at the earth's surface.

It is evident from the data of Chapter 3 that this will require a 2% decrease in direct radiation, which corresponds to a 0.3% decrease in total radiation.

Since the mean value of the aerosol mass that decreases the total radiation by 1% is equal, according to the data presented above, to $0.8 \cdot 10^{-6} \, g/cm^2$, we find that it is necessary to increase the aerosol mass in the northern hemisphere by 600,000 tons.

Different technical means are possible for increasing the aerosol mass. Let us consider one such method, based on the sulphate nature of stratospheric aerosol particles.

To increase the concentration of sulphate particles, we need only deliver a fixed amount of sulphur dioxide to the level of the aerosol layer, which will then be converted into droplets of sulphuric acid. The creation of 600,000 tons of sulphuric acid requires about 400,000 tons of sulphur dioxide, which can be formed by burning 200,000 tons of sulphur.

We will assume that the mean lifetime of aerosol particles in the stratosphere amounts to two years. It will then be necessary to annually deliver 100,000 tons of sulphur into the lower stratosphere in order to maintain the necessary concentration of aerosol particles that consist of pure sulphuric acid. However, since concentrated sulphuric acid is hygroscopic, droplets of it will absorb water vapor in the air, which will increase their mass and decrease their specific weight. This effect will lead to a significant lessening in the amount of reagent necessary for a given decrease in total radiation. Since the exact concentration of sulphuric acid in stratospheric aerosol droplets is unknown, we will take its value as 50%. In this case, about 40,000 tons of sulphur is required. This amounts to less than half the mass of sulphur indicated above due to a decrease in the specific weight of the aerosol as the sulphur dioxide is diluted with water.

Since modern high-altitude aircraft carrying a load of about 15 tons can reach the level of the aerosol layer, this mass of reagent can be transported to the lower stratosphere by several aircraft operating every day equipped with a device for burning sulphur in the atmosphere.

The amount of reagent introduced into the atmosphere by means of this method is negligibly smaller than the mass of matter that enters the atmosphere as a result of man's contemporary economic activity.

This mass is 10^{-4} of that due to man's activity, which, according to contemporary data, constitutes hundreds of millions of tons per year (Inadvertent Climate Modification, 1971).

The amount of reagent finally settling on the earth's surface will be equal in this case to about 0.2 mg of sulphur per square meter per year, which is roughly 10^{-3} of the amount of sulphur that is naturally washed out of the atmosphere by precipitation (Budyko, 1974). Obviously, such amounts are not at all important in environmental pollution.

Though our calculations are highly approximate, refinements in succeeding studies will probably not change the theoretical possibility for climate modification by means of contemporary technical methods of aerosol control in the stratosphere. Such modification will be possible even if it is necessary to increase the amount of reagent ten times, which exceeds the probable error of the influence of aerosol concentration on the radiation flux and of the meteorological solar constant on the air temperature at the earth's surface.

The method of transporting the reagent into the stratosphere by airplanes is probably not the only one; other technical means, including rockets and different types of missiles, are also possible.

It is necessary to investigate the prospect of using other reagents besides sulphur for increasing the mass of stratospheric aerosol. The possibility of lateral effects of the reagent on physical processes in the stratosphere, in particular, on the ozone layer must be studied. This influence is comparatively slight in the range of sulphate aerosol observed under natural conditions. However, it may become substantial when its concentration reaches higher levels or when new reagents not found under natural conditions are introduced into the stratosphere.

The choice of the region where the reagent is scattered is of limited importance since data on the dispersion of products of volcanic eruptions demonstrate that reagent from any point outside the tropical zone rapidly spreads over the entire hemisphere. Circulation in the lower stratosphere can be of importance in selecting optimal regions and periods of time for ejecting the reagent to ensure its most effective use.

These measures of climate modification are intended for preventing or weakening climatic changes that may ensue in several decades as a result of man's economic activity. Such modification, however, is not beyond the capacity of modern technology. In the near future it will apparently be possible to modify the climate, creating climatic changes ten times greater than the fluctuations of the 1920s and 1930s, i.e., producing a drop in global temperature of several degrees.

Effects on such a scale may be necessary in the twenty-first century, when the significant growth in energy production may lead to a substantial increase in the temperature of the lower layers of the atmosphere. Under these conditions, a decrease of the transparency of the stratosphere will prevent undesirable climatic changes.

We should note that it may be feasible to modify the aerosol layer of the stratosphere in the near future, while the influence of man's economic activity on climate will still be comparatively slight.

Modifications that decrease the mean temperature of the lower layers of the atmosphere by several degrees will lead to an enormous change in natural conditions which may be quite unfavorable for man's agricultural activity over the greater part of the globe, and thus in the near future make it impossible to continue this activity.

It is quite realistic to expect that modifications which decrease the mean temperature by several tenths of a degree will be economically feasible. As we noted above, the total amount of precipitation in many middle-latitude regions characterized by insufficient moisture will noticeably decrease with such climatic fluctuations. The ability to prevent such climatic changes can therefore have practical importance.

These modifications can apparently be made without any forecasts of natural changes in the concentration of atmospheric aerosol, using instead current information about this concentration obtained from high-altitude aircraft soundings or from ground actinometric observations. Since, under natural conditions, the mean aerosol concentration in the stratosphere varies comparatively slowly, the rate at which modifications are utilized to maintain this concentration at an optimal level can be controlled.

Let us now discuss the possibility of a climatic change resulting from large numbers of aircraft flights in the stratosphere. This has been dealt with in a collective monograph by American scientists (Man's Influence on the Global Environment, 1970), where it was pointed out that about 500 aircraft flying at altitudes between 18 and 20 km will be used in civil aviation in the years 1985 to 1990. The end-products of the fuel consumption of these aircraft will annually supply 63,000 tons of sulphur dioxide to the lower stratosphere of the entire globe, which corresponds to a mass of sulphate particles of about 100,000 tons.

In considering this question we should bear in mind that, as noted above, a long-term change in radiation will exert a far greater influence on the thermal regime of the atmosphere than a similar change over a short period of time.

Since most of the stratospheric flights will occur in the northern hemisphere, the annual increase in the mass of sulphate particles in this region of the stratosphere can be 70,000–80,000 tons, the total mass of aerosol being at least 200,000 tons.

Stratospheric flights will increase the concentration of particles of the aerosol layer in the northern hemisphere by roughly 400,000 tons if the mean lifetime of these particles is two years; this increase is more than half the amount of aerosol that can cause the type of substantial climatic change considered above. Obviously, this is hardly an insignificant quantity.

Climate modification by altering the sulphur content of fuel in stratospheric aircraft is possible. The above calculation was based on the assumption that fuel with a comparatively low sulphur content of 0.05% will be used in stratospheric flights (Man's Impact on the Global Environment, 1970). This calculation also assumed that the use of fuel with sulphur content up to 0.3% and fuel in which the sulphur content was decreased to 0.01% was possible, though the latter type of fuel is apparently more expensive than fuel with a higher sulphur content.

These data imply that a change in the sulphur content in fuel from 0.01% to 0.30% corresponds to a change in the mass of aerosol particles in the stratosphere of the northern hemisphere from 80,000 to 2,400,000 tons. While the first of these values is not linked to any noticeable climatic change, the second exceeds the change in aerosol content necessary for climate modification. Thus, the prospect of climate modification by varying the type of fuel used by stratospheric aviation becomes promising. In order to maintain stable climatic conditions, it is feasible to use fuel with high sulphur content when the concentration of aerosol particles decreases as a result of natural causes, and to use fuel with low sulphur content when particle concentration is high.

Thus stratospheric flights can be considered a promising element in a system of global climate modification. This problem requires additional study since it is necessary to clarify the influence of stratospheric flights on other elements of the weather conditions, including the ozone content in the stratosphere.

Large numbers of flights in the stratosphere can lead to changes in climatic conditions that must be estimated beforehand.

It is evident from these data that a comparatively slight change in the balance of the mass of stratospheric aerosol can substantially change the air temperature at the earth's surface. It should be noted that such an influence is a few hundred times greater than that of changes exerted by the mass of tropospheric aerosol on the thermal regime of the lower air layers.

This difference is due to a number of causes, for example:

1) the lifetime of a particle in the stratosphere is 50 to 100 times greater than that of an aerosol particle in the troposphere;

2) the stratosphere is essentially lacking in gigantic particles,

which are responsible for about half the mass of tropospheric aerosol and, due to their slight influence on the radiation regime, decrease the optical activity of aerosol in the troposphere to about half;

3) the influence of tropospheric aerosol on the radiation regime is comparatively slight in the presence of clouds, which on the average cover nearly half the earth's surface; this also decreases the optical activity of the aerosol to about half as compared to aerosol in the stratosphere;

4) absorption of short-wave radiation by aerosol in the troposphere partially compensates for the temperature drop at the earth's surface due to a decrease in the radiation flux as the radiation is scattered by aerosol. This effect is chiefly of a local nature in the stratosphere and does not strongly influence the temperature at the earth's surface, since long-wave radiation (on which the temperature of the earth's surface finally depends) departing into outer space remains nearly invariant above the level of the tropopause.

For these reasons, the stratosphere is a favorable zone for climate modifications, and modern advances in stratospheric aviation will help make these modifications highly promising in the near future.

If we agree that it is theoretically possible to produce a noticeable change in the global climate by using a comparatively simple and economical method, it becomes incumbent on us to develop a plan for climate modification that will maintain existing climatic conditions, in spite of the tendency toward a temperature increase due to man's economic activity.

The possibility of using such a method for preventing natural climatic fluctuations leading to a decrease in the rate of the hydrological cycle in regions characterized by insufficient moisture is also of some interest.

The perfection of theories of climate is of great importance for solving these problems, since current simplified theories are inadequate to determine all the possible changes in weather conditions in different regions of the globe that may result from modifications of the aerosol layer of the stratosphere. Obviously, any modifications would be premature before all the consequences can be exactly precalculated.

It is also obvious that the practical realization of this method of climate modification requires the solution of many technical problems associated with the development of optimal reagents and the most feasible methods of ejecting them into the atmosphere.

Thus much preparatory study must be carried out. Methods of preventing unfavorable climate fluctuations should be developed without delay since anthropogenic changes of global climate may significantly intensify over the next decades.

Conclusion

We may conclude from the data set forth in this book that in the modern era the global climate has already been changed to some extent as a result of man's economic activity. These changes are chiefly due to an increase in the mass of aerosol and carbon dioxide in the atmosphere.

Contemporary anthropogenic changes in the global climate have been comparatively small, which is partially explained by the counter-influence of the growth in the concentration of aerosol and carbon dioxide on air temperature. Nevertheless, these changes are of some practical importance, chiefly in connection with the influence of precipitation on agricultural production.

Simple calculations demonstrate that if the contemporary rate of economic development is maintained, anthropogenic changes can rapidly grow and at the end of the twentieth century exceed the natural climate fluctuations that have occurred over the past century.

Climatic changes will intensify from now on under these conditions and in the twenty-first century may become comparable with the natural climatic fluctuations that have occurred over tens of thousands of years of ice ages and interglacial periods. Obviously, such significant climatic changes can enormously influence our planetary environment and many aspects of man's economic activity.

Thus it becomes necessary to predict the anthropogenic climatic changes that will occur under different forms of economic development and also to develop methods of climate modification to prevent undesirable climatic changes. In the light of these problems the importance of studies of climatic changes and, in particular, of the causes of these changes, acquires a new meaning. Until now such studies have been theoretical to a significant extent. However, it is clear that they must become an important tool in planning the development of the national economy.

Refinement and perfection of numerical models of climate theory in order to better predict climatic changes constitute one of the chief lines of research. Since all such existing models are to some degree approximate, it becomes extremely important to verify the models by empirical data. Data on natural climatic changes that have occurred in the past are particularly important in this respect.

Thus the prediction of the climatic conditions of the future is intimately associated with the quantitative explanation of climatic changes that have occurred in the past.

An attempt was made here to use numerical models of climate theory to solve this problem, which resulted in conclusions about the laws for climatic changes in the past and in the future that differ in a number of respects from previously suggested opinions.

The use of more general models of climate theory may allow us to verify these conclusions and to obtain many new data on the laws of climatic changes. Studies along this direction should be greatly expanded in view of the possibility that noticeable changes in the global climate may occur in the near future. The lack of early detailed data on such changes may create enormous difficulties for human society.

The problem of modifying the global climate is of particular importance. The data set forth in this book indicate that climate modification by contemporary technical means can, in theory, be carried out, which increases the urgency of further studies in this field.

The development of methods for predicting the climatic conditions of the future will open horizons for creating numerical models of economic development that take into account the influence of man's economic activity on global climate. Such models can optimize economic planning and eliminate unfavorable changes in environmental conditions with further technological development.

The problem of anthropogenic climatic changes is of international significance and it becomes particularly important in developing large-scale projects of climate modifications. Modification of the global climate will lead to a change in climatic conditions over many countries, the nature of these changes differing in different regions. It has been repeatedly noted by Fedorov (1958) that any large project of climate modification is possible only on the basis of international cooperation.

It appears that there are now grounds for signing an international convention forbidding the implementation of climate modifications that have not been agreed to. Such modifications might be carried out only after being approved by the proper international institutions. This agreement must regulate both measures in the development of climate modifications and those forms of man's economic activity that can lead to inadvertent changes in global climatic conditions.

Summary

The application of physical climatology in studying climatic changes is the main problem presented in this book. It deals with the principal features of modern climate as well as climates of the past. Certain factors that influence climate are particularly discussed

A semiempirical theory of the atmosphere's thermal regime is proposed. This theory enables us to draw conclusions as to the intransivity of the modern climate regime and its high sensitivity to the changes of climate-forming factors.

The semiempirical climate theory is used for a quantitative elucidation of contemporary climatic changes and for a study of the development of the Quaternary glaciation.

Presumably, the climatic variations of the geological past were mainly caused by differences in the distribution of continents and in the composition of the atmosphere.

The stability of the climate depreciated under the influence of the following factors: a drop in temperature occurred in high latitudes and polar ice was formed; this stimulated the development of continental glaciations in high latitudes due to periodic reductions of solar radiation caused by astronomical factors.

Comparatively minor variations in the modern climate are mainly the result of changes in the concentration of stratospheric aerosol particles that influence solar radiation flux. These changes are brought about by volcanic activity.

The book also deals with the effect of climatic changes on biological processes including the evolution of living organisms.

In the last decades global climate has been affected by anthropogenic factors which include an increase of carbon dioxide and aerosol concentrations in the atmosphere.

This influence is rapidly augmenting and may in the near future result in drastic changes of climate of great practical significance to man. It is very essential, therefore, to develop methods for controlling climate modifications. One such method is to maintain a certain level of aerosol concentration in the stratosphere.

Bibliography

Adhèmar, J. F., Les rèvolutions de la mar. Déluges périodiques, Paris, 184 pp., 1842.
Ahlman, H. W., Glacier variations and climatic fluctuations, Amer. Geogr. Soc., Ser. 3, N.Y., 51 pp., 1953.
Albrecht, F., Doe Aktiongebiete des Wasser- und Wärmehaushaltes der Erdoberfläche, Z. Meteorol., 1, 4–5, 97–109, 1947.
Allen, C. W., Solar radiation, Quart. J. Roy. Meteorol. Soc., 84, 362, 307–318, 1958.
Alyea, F. N., Numerical simulation of an ice age paleoclimate, Atmos. Sci. Pap., No. 193, Colorado State Univ., 120 pp., 1972.
Angström, A., Atmospheric turbidity, global illumination and planetary albedo of the earth, Tellus, 14, 4, 435–450, 1962.
Angström, A., The solar constant and the temperature of the Earth, Progress in Oceanography, V. 3, Permagon Press, Oxford, pp. 1–5, 1965.
Angström, A., Apparent solar constant variations and their relation to the variability of atmospheric transmission, Tellus, 21, 2, 205–218, 1969.
Arrhenius, S., On the influence of the carbonic acid in the air upon the temperature of the ground, Philos. Mag., 41, 237–275, 1896.
Arrhenius, S., Lehrbuch der kosmischen Physik, Hirzel, Leipzig, 1026 S., 1903.
Atlas of Heat Balance (in Russian), ed. M. I. Budyko, Leningrad, Izd. G.G.O., 41 pp., 1955.
Atlas of Heat Balance of the World (in Russian), ed. M.I. Budyko, Moscow, Mezhduvedomstvennyi geofizicheskii komitet, 69 pp., 1963.
Atwater, M. A., Planetary albedo changes due to aerosol, Science, 170, 3953, 64–66, 1970.
Auer, V., The Pleistocene of Fuego-Patagonia, Part I, The ice age and interglacial ages, Suomal. Tiedeakat. Toim. Helsinki, Ser. A, III, Geologica-Geographica, pp. 50, 239, 1956.
Axelrod, D. I., Quaternary extinctions of large mammals, Univ. of Calif. Publ. Geol. Sci., 74, 42 pp., 1967.
Axelrod, D. I., and H. P. Baily, Cretaceous dinosaur extinction, Evolution, 22, 3, 595–611, 1968.
Barrett, E., Depletion of short-wave irradiance at the ground by particles suspended in the atmosphere, Solar energy, 13, 3, 323–337, 1971.
Bellaire, A., and R. Garrington, The World of Reptiles, London, 153 pp., 1966.

Berger, A. L., Théorie astronomique des paleoklimats, V. 1, 197 pp., V. 2, 108 pp., Univ. catholique de Louvain, Louvain, 1973.
Berlyand, M. E., and K. Ya. Kondrat'ev, Cities and Climate of the Planet (in Russian), Leningrad, Gidrometeoizdat, 40 pp., 1972.
Berlyand, M. E., et al., Complex study of the perculiarities of the meteorological regime in a large city (in Russian), Meteorol. Gidrol., 1, 14–23, 1974.
Bibikov, S. N., Certain aspects of the paleoeconomic model of the Paleolite (in Russian), Sov. Archeol., 4, 5–22, 1969.
Blinova, E. N., On the mean annual temperature distribution in the earth's atmosphere with consideration of continents and oceans (in Russian), Izv. AN USSR, Ser. Geogr. Geofiz., 11, 1, 3–13, 1947.
Bogert, C. M., How reptiles regulate their body temperature?, Sci. American, 200, 4, 105–108, 1959.
Bolin, B., and W. Bischof, Variation of the carbon dioxide content of the atmosphere in the northern hemisphere, Tellus, 22, 4, 431–442, 1970.
Bolin, B., and C. D. Keeling, Large-scale atmospheric mixing as deduced from the seasonal and meridional variations of carbon dioxide, J. Geophys. Res., 68, 13, 3899–3920, 1963.
Borzenkova, I. I., On the possible influence of volcanic dust on the radiation and thermal regime (in Russian), Tr. GGO, Vyp. 307, 36–42, 1974.
Borzenkova, I. I., K. Ya. Vinnikov, L. P. Spirina, and D. I. Stekhnovskiy, Change of air temperature of the Northern Hemisphere for 1881–1975 (in Russian), Meteorol. Gidrol., 7, 27–35, 1976.
Bowen, R., Paleotemperature Analysis, Elsevier, 265 pp., Amsterdam, London, New York, 1966.
Braidwood, R. I., and C. A. Reed, The achievement and early consequences of food production, Cold Spring Harbor Symp. Quant. Biol., V. 22, Population Studies, Animal Ecology and Demography, pp. 19–31, New York, 1957.
Brooks, C. E. P., Climate Through the Ages, 2nd ed., Ernst Benn, London, 1950.
Bryson, R. A., All other factors being constant . . . , Weatherwise, 21, 2, 56–62, 1968.
Bryson, R. A., Climate modification by air pollution, Report 1, Univ. of Wisconsin, Madison, 16 pp., 1972.
Budyko, M. I., Heat Balance of the Earth's Surface (in Russian), Leningrad, Gidrometeoizdat, 255 pp., 1956.
Budyko, M. I., On the thermal zonality of the earth (in Russian), Meteorol. Gidrol., 11, 7–14, 1961.
Budyko, M. I., Polar ice and climate (in Russian), Izv. AN USSR, Ser. Geogr., 6, 3–10, 1962.
Budyko, M. I., Certain methods of influencing the climate (in Russian), Meteorol. Gidrol., 2, 3–8, 1962a.
Budyko, M. I., Climatic changes (in Russian), Meteorol. Gidrol., 11, 18–27, 1967.
Budyko, M. I., Reasons for the extinction of certain animals at the end of the Pleistocene (in Russian), Izv. AN SSR, Ser. Geogr. 2, 28–36, 1967a.
Budyko, M. I., On the origin of ice ages (in Russian), Meteorol. Gidrol, 11, 3–12, 1968.

Budyko, M. I., Climatic Changes (in Russian), 35 pp., Gidrometeoizdat, Leningrad, 1969.
Budyko, M. I., The effect of solar radiation variations on the climate of the earth, Tellus, 21, 5, 611–619, 1969a.
Budyko, M. I., Comments on "A global climatic model based on the energy balance of the earth-atmosphere system", J. Appl. Meteorol., 9, 2, 310–311, 1970.
Budyko, M. I., The Climate and Life (in Russian), 470 pp., Gidrometeoizdat, Leningrad, 1971.
Budyko, M. I., Comments on "Intransitive model of the earth-atmospheric-ocean system", J. Appl. Meteorol., 11, 7, 1150–1151, 1972.
Budyko, M. I., The future climate, EOS, 53, 10, 868–874, 1972a.
Budyko, M. I., Man's Influence on Climate (in Russian), 47 pp., Gidrometeoizdat, Leningrad, 1972b.
Budyko, M. I. Atmospheric Carbon Dioxide and the Climate (in Russian), 32 pp., Gidrometeoizdat, Leningrad, 1973.
Budyko, M. I., Man and the biosphere (in Russian), Vopr. Fil., 1, 61–70, 1973a.
Budyko, M. I., Method of influencing the climate (in Russian), Meteorol. Gidrol. 2, 91–97, 1974.
Budyko, M. I., Climate of the future (in Russian), Meteorol. Gidrol., 9, 3–15, 1974a.
Budyko, M. I., The dependence of the mean air temperature on the change of solar radiation (in Russian), Meteorol. Gidrol. 10, 3–10, 1975.
Budyko, M. I., O. A. Drozdov, and M. I. Yudin, The influence of man's economic activity on the climate, in Contemporary Problems of Climatology (in Russian), pp. 435–448, Gidrometeoizdat, Leningrad, 1966.
Budyko, M. I., and Z. I. Pivovarova, Effect of volcanic eruptions on the influx of solar radiation at the earth's surface (in Russian), Meteorol. Gidrol., 10, 3–7, 1967.
Budyko, M. I., and M. A. Vasishcheva, Effect of astronomical factors on Quaternary glaciations (in Russian), Meteorol Gidrol., 6, 37–47, 1971.
Budyko, M. I., and K. Ya. Vinnikov, Contemporary Climatic Changes (in Russian), Meteorol. Gidrol., 9, 3–13, 1973.
Budyko, M. I., and K. Ya. Vinnikov, Global warming (in Russian), Meteorol. Gidrol., 7, 16–25, 1976.
Budyko, M. I., and M. I. Yudin, On the level fluctuations in drainless lakes (in Russian), Meteorol. Gidrol., 8, 15–19, 1960.
Budyko, M. I., et al., Changes in the Climate in Connection with the Plan of Nature Transformation (in Russian), 206 p. Gidrometeoizdat, Leningrad, 1952.
Budyko, M. I., et al., Prospects of changing the global climate (in Russian), Izv. AN SSSR, Ser. Geogr., 2, 11–23, 1974.
Burdecki, F., Phenomena after volcanic eruptions, Weather, 19, 4, 113–114, 1964.
Butzer, K. W. Environment and Archeology, 524 pp., Methuen, London, 1964.
Cadle, R. D., Composition of the stratospheric "Sulfate layer", Proceedings of

the survey conference Feb. 15-16, 1972, U. Department of Transportation, 1972.

Callender, G. S., The artificial production of carbon dioxide and its influence on temperature, Quart. J. Roy. Meteorol. Soc., 64, 27, 223-240, 1938.

Cess, R. D., Climatic change: an appraisal of atmospheric feedback mechanisms employing zonal climatology, J. Atmos. Sci. (in print), 1976.

Chamberlin, T. C., A group of hypotheses bearing on climatic changes, J. Geology, 5, 653-683, 1897.

Chamberlin, T. C., The influence of great epochs of limestone formation upon the constitution of the atmosphere, J. Geology, 6, 609-621, 1898.

Chamberlin, T. C., An attempt to frame a working hypothesis of the cause of glacial periods on an atmospheric basis, J. Geology, 7, 545-584, 1899.

Charlson, R. J., and M. J. Pilat, Climate: The influence of aerosols, J. Appl. Meteorol., 8, 6, 1001-1002, 1969.

Chizhov, O. P., and A. M. Tareeva, Possibilities of estimating the ice regime of the Arctic basin and its variation, in Materials of glaciological studies (in Russian), 15, 57-73, 1969.

Colbert, E. N., Evolution of the Vertebrates, 479 pp., John Wiley & Sons, N.Y., 1958.

Colbert, E. N., The Age of Reptiles, 228 pp., Weidenfeld & Nicolson, London, 1965.

Darwin, Ch., The Origin of Species by Means of Natural Selection, Murray, London, 1859.

Darwin, Ch., The Descent of Man and Selection in Relation to Sex, Murray, London, 1871.

Davitaya, F. F., On the possible influence of atmospheric dust on the melting of glaciers and warming of the climate (in Russian), Izv. AN SSSR, Ser. Georgr. 2, 3-23, 1965.

Davitaya, F. F., History of the atmosphere and the dynamics of its gaseous composition (in Russian), Meteorol. Gidrol., 2, 21-28, 1971.

Donn, W. L., and D. M. Shaw, The heat budgets of an ice-free and ice-covered Arctic Ocean, J. Geophys. Res., 71, 4, 1087-1093, 1966.

Dorodnitsyn, A. A., B. I. Izvekov, and M. E. Shvets, Mathematical theory of general circulation (in Russian), Meteorol. Gidrol., 4, 32-42, 1939.

Drozdov, O. A., and A. S. Grigor'eva, Moisture Circulation in the Atmosphere (in Russian), 314 pp., Gidrometeoizdat, Leningrad, 1963.

Drozdov, O. A., and A. S. Grigor'eva, Long-term Cyclic Fluctuations of Atmospheric Precipitation over the USSR (in Russian), 158 pp., Gidrometeoizdat, Leningrad, 1971.

Drozdov, O. A., and T. V. Pokrovskaya, On the role of random variations of the water balance in the fluctuations of certain lakes (in Russian), Meteorol. Gidrol., 8, 43-48, 1961.

Dwayr, H. A., and T. Peterson, Time-dependent energy modeling, J. Appl. Meteorol., 12, 36-42, 1973.

Dyer, A. J., and B. B. Hicks, Stratospheric transport of volcanic dust inferred from solar radiation measurements, Nature, 208, 5006, 131-133, 1965.

Dyer, A. J., and B. B. Hicks, Global spread of volcanic dust from the Bali eruption of 1963, Quart. J. Roy. Meteorol. Soc., 94, 402, 545–554, 1968.

Elton, Ch., Animal Ecology and Evolution, 96 pp., Clarendon Press, Oxford, London, 1930.

Emiliani, C., Pleistocene temperature, J. Geol., 63, 6, 538–578, 1955.

Emiliani, C., Isotopic paleotemperatures, Science, 154, 851–857 1966.

Ensor, D. S., et al , Influence of atmospheric aerosol on albedo J. Appl. Meteorol., 10, 6, 1303–1306, 1972.

Erikson, E., Air-ocean-ice cap interactions in relation to climatic fluctuations and glaciation cycles, Causes of climatic change INQUA, USA, 1968.

Ewing, D. M., and W. L. Donn, A theory of ice ages, I, Science, 123, 1061–1066, 1956.

Ewing, D. M., and W. L. Donn, A theory of ice ages, II, Science, 127, 1159–1162, 1958.

Faegre, A., An intransitive model of the Earth-atmosphere-ocean system, J. Appl. Meteorol., 11, 1, 4–6, 1972.

Fairbridge, R. W., Ice-age theory, in The Encyclopaedia of Atmospheric Sciences and Astrogeology, pp. 462–474, Reinhold Publ. Corp., New York-Amsterdam-London, 1967.

Fedorov, E. K., Man's influences on meteorological processes (in Russian), Vopr. Fil., 4, 137–144, 1958.

Fedorov, E. K., Interaction of Mankind and Nature (in Russian), 88 pp., Gidrometeoizdat, Leningrad, 1972.

Fletcher, J. O., The Arctic heat budget and atmospheric circulation, Proc. of the Symposium on the Arctic heat budget and atmospheric circulation, pp. 23–32, Rand Corp., Santa Monica, Calif., 1966.

Flint, R. F., Glacial Geology and the Pleistocene Epoch, New York-London, 1963.

Flohn, H., Zur meteorologische Interpretation der Pleistozänen-Kilmaschwänkungen, Eiszeitalter und Gegenwart, 14, 153–160, 1963.

Flohn, H., Grundfragen der Paläoklimatologie in Lichte einer theoretischen Klimatologie, Geologische Rundschau, 54, 5, 504–515, 1964.

Flohn, H., Ein geophysikalisches Eiszeit-Modell, Eiszeitalter und Gegenwart, 20, 204–231, 1969.

Flowers, E. C., and H. J. Viebrock, Solar radiation – an anomalous decrease of direct solar radiation, Science, 148, 3669, 493–494, 1965.

Forrester, J. W., World Dynamics, 142 pp., Wright-Allen Press, Cambridge, Mass., 1971.

Fuchs, V. E., and T. T. Patterson, The relation of volcanity and orogeny to climatic change, Geol. Mag., 84, 6, 321–333, 1947.

Gaevskaya, G. N., Absorption of short-wave radiation by aerosol (in Russian), Probl. Fis. Atmos., 10, 86–91, Izd. LGU, Leningrad, 1972.

Gandin, L. S., B. M. Il'in, and L. V Rukhovets, On the influence of variations of external parameters on the atmospheric thermal regime (in Russian), Tr. GGO, 315, 21–38, 1973.

George, W., Biologist Philosopher, 320 pp., Abelard-Schuman, London-Toronto-New York, 1964.
Gerasimov, I. P., Nature and evolution of primeval society (in Russian), Izv. AN USSR, Ser. Geogr. 1, 5-8, 1970.
Gerasimov, I. P., and M. I. Budyko, Actual problems of man's interaction with nature (in Russian), Kommunist, 10, 79-91, 1974.
Gerasimov, I. P. and K. K. Markov, The ice Age and Territory of the USSR, 462 pp., Izd. AN USSR, Moscow-Leningrad, 1939.
Gol'tsberg, I. A., Expedition for the study of atmospheric turbulence in forest strips (in Russian), Tr. GGO, 29, 5-10, 1952.
Gordon, H. B., and D. H. Davis. The effect of changes in solar radiation on climate, Quart. J. Roy. Meteorol. Soc., 100, 423, 123-126, 1974.
Grigor'ev, A. A. The Sub-Arctic (in Russian), 233, pp., Geografgiz, Moscow, 1956.
Grigor'ev, A. A., and M. I. Budyko, On the periodical law of geographical zonality (in Russian), AN USSR, 110, 1, 129-132, 1956.
Grobecker, A. J., Assessment of Climatic Changes due to Flights in the Stratosphere, 16 pp., Dept. Transportation, Washington, D. C., 1973.
Hantel, M., Polar boundary conditions in zonally averaged global climatic models, J. Appl. Meteorol., 13, 752-759, 1974.
Held, I. M., and J. M. Suarez, Simple albedo feedback models of the ice-caps, Tellus, 26, 6, 613-630, 1974.
Hoffman, E., Der Dritte Karbonstratigraphische Kongress in Heerlen, Phyton, 3, 3-4, 193-209, 1951.
Holloway, J. L., and S. Manabe, A global general circulation model with hydrology and mountains, Mon. Weather Rev., 99, 5, 355-370, 1971.
Humphreys, W. J., Volcanic dust and other factors in the production of climatic changes and their possible relation to ice age, J. Franklin Inst., 1976, 131 pp., 1913.
Humphreys, W. J., Physics of the Air, 2d ed., 654 pp., McGraw-Hill Book Co., New York, 1929.
Huntington, E., and S. S. Visher, Climatic Changes. Their Nature and Causes, 329 pp., Yale Univ. Press, New Haven, Conn., 1922.
Inadvertent Climate Modification (SMIC). The MIT Press, Cambridge, Mass., and London, 306, 1971.
Junge, C., Atmospheric Chemistry and Radioactivity, Academic Press, New York, 1963.
Kagan, R. L., and K. Ya. Vinnikov, On the assumption of heat inflow during numerical experiments with a thermotropic model (in Russian), Tr. GGO, 256, 98-107, 1970.
Kalesnik, S. V., Principles of Physical Geography (in Russian), Uchpedgiz, Moscow-Leningrad, 1947.
Kalitin, N. N., On the time of the onset of the optical anomaly in 1912 (in Russian), Izv. GFO, 1, 11-17, 1920.
Kaplan, L. D., The influence of carbon-dioxide variation on the atmospheric heat balance, Tellus, 12, 2, 209-216, 1960.

Karol', I. L., Radioactive Isotopes and Global Transfer in the Atmosphere (in Russian), 365 pp., Gidrometeoizdat, Leningrad, 1972.
Karol', I. L., The size of radioactive aerosol and its transfer in the troposphere and stratosphere (in Russian), Meteorol. Gidrol., 1, 28–38, 1973.
Kellogg, W. W., Mankind as a Factor in Climate Change. The Energy Question, Univ. of Toronto Press, 1974.
Kimball, H. H., Volcanic eruptions and solar radiation intensities, Mon. Weather Rev., 46, 8, 355–356, 1918.
Kochin, N. E., Compiling model for the zonal circulation of the atmosphere (in Russian), Tr. GGO, 10, 3–27, 1936.
Kondrat'ev, K. Ya., and N. Y. Niilisk, On the question of carbon dioxide heat radiation in the atmosphere, Geophys. Pura Appl., 46, 216–230, 1960.
Kondrat'ev, K. Ya., et al., The Influence of Aerosol on the Transfer of Radiational; Possible Climatic Consequences (in Russian), 266 pp., Izd. LGU, Leningrad, 1973.
Köppen, W., and A. Wegener, Die Klimate der geologischen Vorzeit, 256 S., Berlin, 1924.
Kurten, B., Evolution in Geological Time, in Ideas in Modern Biology, ed. by J. A. Moore, 563 pp., Natural History Press, Garden City, N.Y., 1965.
Kurten, B., Time and hominid brain size, Comment Biol., 36, 1–8, Helsinki, 1971.
Kuznetsova, L. P., and V. Ya. Sharova, Amount of Precipitation (in Russian), God. Fiz. Geogr. Atlas Mira, 42–43, 1964.
Laek, D., The Natural Regulation of Animal Numbers, Oxford, 1954.
Lamb, H. H., The role of atmosphere and oceans in relation to climatic changes and the growth of ice sheets on land, Problems in Paleoclimatology, 332–348 pp. Interscience Publishers, London-New York-Sydney, 1964.
Lamb, H. H., The Changing Climate, 236 pp., Methuen, London, 1966.
Lamb, H. H., Volcanic dust in the atmosphere . . . Phil. Trans., Roy. Soc. of London, Appl. Math. Phys. Sci., 266, 1178, 425–533, 1970.
Lamb, H. H., Climate, Present, Past and Future, V. 1, 612 pp., Methuen, London, 1973.
Lamb, H. H., Climatic change and foresight in agriculture: the possibilities of long-term weather advice, Outlook on Agriculture, 7, 5, 203–210, 1973.
Lamb, H. H., The Current Trend of World Climate – a Report on the Early 1970's and a Perspective, Univ. of East Anglia, Norwich, 28 pp., 1974.
Lamb, H. H., S. A. Malberg, and J. M. Colebrook, Climatic reversal in the northern North Atlantic, Nature, 256, 5517, 479, 1975.
Landsberg, H. E., The Climate of Towns. Man's Role in Changing the Face of the Earth, pp. 584–606, Univ. Chicago Press 1956.
Landsberg, H. E., Note on the recent climatic fluctuation in the United States, J. Geophys. Res., 65, 5, 1519–1525, 1960.
Landsberg, H. E., Man-made climatic changes, Science, 170, 1264–1265, 1970.
Landsberg, H. E., and T. N. Maisel, Micrometeorological observations in an area of urban growth, Boundary-Layer Meteorol. 2, 365–370, 1972.
Landsberg, H. E., C. S. Yu, and L. Huang, Preliminary Reconstruction of a

Long Time Series of Climatic Data for the Eastern United States, 20 pp., Univ. of Maryland, Inst. of Fluid Dynamics and Appl. Math. Techn., Note BN-57, 1968.

Lieth, H., The role of vegetation in the carbon content of the atmosphere, J. Geophys. Res., 68, 13, 3887–3898, 1963.

Lorenz, E. N., Climatic determinism, Meteorol. monogr., 8, 30, 1–3, 1968.

Lukashevich, I. D., On the causes of the ice age (in Russian), Priroda, 959–979, St. Petersburg, July–August, 1915.

Lyell, C., Principles of Geology, V. 1, 511 pp., 1830; V. 2, 330 pp., 1832; V. 3, 398 pp., 1833, London, 1830–1833.

Manabe, S., The Dependence of Atmospheric Temperature on the Concentration of Carbon Dioxide. Global Effects of Environmental Pollution, 218 pp., Reidel Publ. Co., Dordrecht, Holland, 1970.

Manabe, S., and K. Bryan, Climate calculation with a combined ocean-atmosphere model, J. Atmos. Sci., 26, 786–789, 1969.

Manabe, S., and R. T. Wetherald, Thermal equilibrium of the atmosphere with a given distribution of relative humidity, J. Atmos. Sci., 24, 241–259, 1967.

Manabe, S., and R. T. Wetherald, The effects of doubling the CO_2 concentration on the climate of a general circulation model, J. Atmos. Sci., 32, 1, 3–15, 1975.

Man's Impact on the Global Environment (SCEP), MIT Press, Cambridge, Mass., and London, 318 pp., 1970.

Markov, K. K., Essays on the Geography of the Quaternary Period (in Russian), 347 pp., Geografgiz, Moscow, 1955.

Markov, K. K., Paleography (in Russian), 2nd ed., 268 pp., Izd. MGU, Moscow, 1960.

Martin, P. S., Pleistocene ecology and biography of North America, Amer. Assoc. Adv. Sci. Publ., 51, 509 pp., 1958.

Martin, P. S., Africa and Pleistocene overkill, Nature, 212, 5060, 339–342, 1966.

Martin, P. S., Prehistoric overkill. Pleistocene extinctions, Proc. III Congr. Intern. Assoc. Quart. Res., 75–120, Yale Univ. Press, New York-London, 1967.

Martin, P. S., The discovery of America, Science, 179, 969–974, 1973.

Mathew, W. D., Climate and Evolution, 2nd ed., 1915, Spec. Publ., New York Acad. Sci. 1, 223 pp., 1939.

Mayr, E., Animal Species and Evolution, 797 pp., Cambridge, Mass., 1963.

McCormic R. A., and J. H. Ludwig, Climate modification by atmospheric aerosols, Science, 156, 1358–1359, 1967.

Meadows, D. H., et al., Limits to Growth, 205 pp., Universe Book, New York, 1972.

Milankovich, M., Théorie mathématique des phénoménes thermiques produits par la radiation solaire, 339 pp., Paris, 1920.

Milankovich, M., Mathematische Klimalehre und astronomische Theory der Klimaschwankungen, Köppen-Geiger, Handb. Klimat., I, A., 176 pp., Berlin, 1930.

Milankovich, M., Kanon der Erdbestrahlung und seine Anwendung am Eiszeitproblem, 133, 633 pp., Royal Serbian Academy, 1941.

Mintz, Y., Very long-term global integration of the primitive equation of atmospheric motion, WMO Technical Note, 66, 141–161, 1965.

Mitchell, J. M., Recent secular changes of global temperature, Ann. N. Y. Acad. Sci., 95(1), 235–250, 1961.

Mitchell, J. M., On the world-wide pattern of secular temperature changes, UNESCO, Arid Zone Research, 20, 161–180, Paris, 1963.

Mitchell, J. M., The solar inconstant, Proc. Seminar on Possible Responses of Weather Phenomena to Variable Extraterrestrial Influences, NCAR Techn. Note, TN-8, 155–174, Boulder, Colorado, 1965.

Mitchell, J. M., Theoretical paleoclimatology. The Quaternary of United States, 881–901, Univ. Press, Princeton, 1965a.

Mitchell, J. M., Concluding remarks. Causes of climatic changes, 155–159, INQUA, USA, 1968.

Mitchell, J. M., The effect of atmospheric aerosols on climate with special reference to temperature near the Earth's surface, J. Appl. Meteorol., 10, 4, 703–714, 1971.

Möller, F., Vierteljahreskarten des Niederschlags für die ganze Erde, Pettermanns Geogr. Mitt., 95, 1, 1–7, 1951.

Möller, F., On the influence of changes in the CO_2 concentration in air on the radiation balance of the Earth's surface and on climate, J. Geophys. Res., 68, 13, 3877–3886, 1963.

Müller, D., Kreislauf des Kohlenstoffers, Handbuch, die Pflanzenphysiologie, 12, 2, 934–948, Berlin, 1960.

Newell, R. L., The global circulation of atmospheric pollutants, Scientific American, 224, 1, 32–42, 1971.

Newmen, J., and A. Cohen, Climatic effects of aerosol layers in relation to solar radiation, J. Appl. Meteorol., 11, 4, 651–657, 1972.

North, G. R., Analytical solution to a simple climate model with diffusive heat transport, J. Atmos. Sci., 32, 7, 1301–1307, 1975.

North, G. R., Theory of energy balance climate models, J. Atmos. Sci., 32, 11, 2033–2044, 1975a.

Opik, E. J., On the causes of the paleoclimatic variations and of the Ice Ages in particular, J. Glaciol., 2, 13, 213–219, 1953.

Opik, E. J., Solar variability and paleoclimatic changes, Irish Astron. J., 5, 4, 5–97, 1958.

Pidoplichko, I. G., Contemporary problems and the study of the history and environmental conditions of faunas in natural conditions and faunas in the past (in Russian), 1, 152 pp., Izd. AN Ukr. SSR, Kiev, 1963.

Pivovarova, Z. I., Long-term variations of the solar radiation intensity based on observations at actinometric stations (in Russian), Tr. GGO, 233, 17–29, 1968.

Plass, G. N., The carbon dioxide theory of climate change, Tellus, 8, 2, 140–154, 1956.

Pokrovskaya, T. V., On the radiation factors of climatic variations (in Russian), Meteorol. Gidrol., 10, 31–38, 1971.
Rakipova, L. R., Thermal Regime of the Atmosphere (in Russian), Gidrometeoizdat, 183 pp., Leningrad, 1957.
Rakipova, L. R., Climatic changes as a result of the modification of the Arctic ice (in Russian), Meteorol. Gidrol., 9, 28–30, 1962.
Rakipova, L. P., Changes of the zonal distribution of atmospheric temperature as a result of active climatic modifications in Contemporary Problems of Climatology (in Russian), 358–384, Gidrometeoizdat, Leningrad, 1966.
Rakipova, L. R., and O. N. Vishnyakova, The dependence of the atmospheric thermal regime on the carbon dioxide concentration (in Russian), Meteorol. Gidrol., 5, 23–32, 1973.
Ramsay, W., Orogenesis und Klima. Ofversigt of Finska Vetensk. Soc. Förh., 52 A, 11, 1910.
Raschke, E., F. Möller, and W. Bandeen, The radiation balance of the Earth-atmosphere system over both polar regions obtained from radiation measurements of the Numbus II meteorological satellite, Sveriges Meteorol. och Hydrol. Inst. Medellanden, Ser. B, 28, 104, 1968.
Raschke, E., et al., The radiation balance of the Earth-atmosphere system from Nimbus III radiation measurements, NASA Technical Note, D-7242, 73 pp., Washington, 1973.
Rasool, S. I., and S. H. Schneider, Atmospheric carbon dioxide and aerosol effect of large increases on global climate, Science, 173, 138–141, 1971.
Robinson, C. D., Surface measurements of solar and terrestrial radiation during the IGY and IGG, Ann. Intern. Geophys. Year, 32, 17–61, 1964.
Romanov, V. V., Hydrophysics of Swamps (in Russian), 359 pp., Gidrometeoizdat, Leningrad, 1961.
Romer, A. S., Vertebrate Paleontology, 687 pp., Univ. of Chicago Press, Chicago, 1945.
Ronov, A. B., On the post-Cambrian geochemical history of the atmosphere and hydrosphere (in Russian), Geokhim., 5, 397–409, 1959.
Ronov, A. B., General tendencies in the evolution of the composition of the Earth's crust, ocean and atmosphere, Geokhim., 8, 715–743, 1964.
Ronov, A. B., Evolution of rock composition and geochemical processes in the Earth's lithosphere (in Russian), Geokhim., 2, 137–147, 1972.
Rosenberg, G. V., Optical studies of atmospheric aerosol (in Russian), Fiz. Nauk, 95, 1, 159–208, 1968.
Rubinshtein, E. S., On the problem of climatic changes (in Russian), Tr. NIU, GUGMS, 1, 22, 3–83, 1946.
Rubinshtein, E. S., The Structure of Fluctuations of Air Temperature in the Northern Hemisphere (in Russian), 34 pp., Gidrometeoizdat, Leningrad, 1973.
Rubinshtein, E. S., and L. G. Polozova, Contemporary Climatic Changes (in Russian), 268 pp., Gidrometeoizdat, Leningrad, 1966.
Rukhin, L. B., Problems of the origin of continental glaciations (in Russian), Izv. VGO, 1, 25–38, 1958.

Rukhin, L. B., Principles of General Paleogeography, 2nd ed., 628 pp., Gostoptekhizdat, Leningrad, 1962.
Rutten, M. G., The Origin of Life by Natural Causes, Elsevier Publ. Co., Amsterdam-London-New York, 1971.
Saks, V. N., The Quaternary Period in the Soviet Arctic (in Russian), 2nd ed., Vodtransizdat, 267 pp., Leningrad-Moscow, 1953.
Saks, V. N., Geological history of the Arctic Ocean during the Mesozoic era (in Russian), Intern. Geol. Congr. XXI session, Reports of Soviet geologists, pp. 108–123, Gosgeoltekhizdat, Moscow, 1960.
Saltzman, B., and A. D. Vernekar, Note of the effect of Earth orbital radiation variations on climate, J. Geophys. Res., 76, 18, 4195–4197, 1971.
Sarasin, P., and F. Sarasin, La température de la période glaciare, Verh. Naturforsch. Ges., 13, 603–615, Basel, 1901.
Savinov, S. I., Maximum values of the solar radiation intensity according to observations in Pavlovsk beginning in 1892 (in Russian) Izv. AN USSR, 6, 7, 12, 707–720, 1913.
Sawyer, J. S., Notes on the response of the general circulation to changes in the solar constant. Changes of climate, Proc. Rome Symp. UNESCO, 333–338, 1963.
Sawyer, J. S., Possible variations of the general circulation of the atmosphere. World climate from 8000 to 0 B.C., Proc. Intern. Symp. 18–19 April, 1966, Roy. Meteorol. Soc., 218–229, 1966.
Schell, J. J., On mathematical and "natural" models for the study of climate changes, J. Appl. Meteorol., 10, 6, 1344–1346, 1971.
Schneider, S. H., Cloudiness as a global climatic feedback mechanism: the effects on the radiation balance and surface temperature of variations in cloudiness, J. Atmos. Sci., 29, 1413–1422, 1972.
Schneider, S. H., and Tzvi Gal-Chen, Numerical experiments in climate stability, J. Geophys. Res., 78, 27, 6182–6194, 1973.
Schwarzbach M., Das Klima der Vörzeit. Eine Einführung in die Paläoklimatologie, 211 S., Stuttgart, 1950.
Schwarzbach, M., Neue Eiszeithypotesen, Eiszeitalter and Gegenw., 19, 250–261, 1961.
Schwarzschild, M., Structure and Evolutions of the Stars, 296 pp., Princeton Univ. Press, Princeton, 1958.
Sellers, W. D., Physical Climatology, 272 pp., Univ. Chicago Press, 1965.
Sellers, W. D., A global climatic model based on the energy balance of the Earth-atmosphere system, J. Appl. Meteorol., 8, 3, 392–400, 1969.
Sellers, W. D., The effect of changes in the Earth's obliquity on the distribution of mean annual sea-level temperature, J. Appl. Meteorol., 9, 960–961, 1970.
Sellers, W. D., A new global climatic model, J. Appl. Meteorol., 12, 2, 241–254, 1973.
Sellers, W. D., The effect of solar constant variation on climate modeling, Proc. Workshop: The Solar Constant and the Earth's Atmosphere, Calif. Inst. Technol. p. 206–209, 1975.
Severtsev, S. A., Population Dynamics and the Adaptation Evolution of Animals (in Russian), 316 pp., Izd. AN SSSR, 1941.

Sharaf, Sh. G., and N. A. Budnikova, Secular variations of the elements of the Earth's orbit and the astronomical theory of climatic changes (in Russian), Tr. Inst. Teor. Astron., 14, 48–85, 1969.
Shifrin, K. S., and I. N. Minin, On the theory of nonhorizontal visibility (in Russian), Tr. GGO, 68, 5–75, 1957.
Shifrin, K. S., and N. F. Pyatovskaya, Tables of the Visibility Range and Daylight Sky Brightness (in Russian), 211 pp., Gidrometeoizdat, Leningrad, 1959.
Shmal'gauzen, I. II., Evolutionary Factors (in Russian), 396 pp., Izd. AN USSR, Moscow, 1946.
Shnitnikov, A. V., Intrasecular Variability of the General Humidity Components (in Russian), 245 pp., Nauka, Leningrad, 1969.
Shuleikin, V. V., Physics of the Sea (in Russian), 1083 pp., Izd. AN USSR, Moscow-Leningrad, 1941, 1953, 1968.
Shvets, M. E., et al., Numerical model of general atmospheric circulation over a hemisphere (in Russian), Tr. GGO, 256, 3–44, 1970.
Simpson, G. C., Past climates, Quart. J. Roy. Meteorol. Soc., 53, 221, 213–226, 1927.
Simpson, G. C., Ice Ages, Nature, 141, 591–598, 1938.
Simpson, G. C., Tempo and Mode of Evolution, New York, 1944.
Sinitsyn, V. M., Introduction to Paleoclimatory (in Russian), 232 pp., Nedra, Leningrad, 1967.
Smagorinsky, J., General circulation experiments with the primitive equations. 1. The basic experiment, Mon. Weather Rev., 91, 99–164, 1963.
Smagorinsky, J., Global atmospheric modeling and the Numerical simulation of Climate. Weather Modification, ed. W. N. Hess, John Wiley & Sons, New York, 1974.
Smagorinsky, J., S. Manabe, and I. L. Holloway, Numerical results from nine-level general circulation model of the atmosphere, Mon. Weather Rev., 93, 727–768, 1965.
Spirina, L. P., On the influence of volcanic dust on the thermal regime of the northern hemisphere (in Russian), Meteorol. Gidrol., 10, 38–45, 1971.
Strakhov, N. M., Historical Geology (in Russian), 2, 428 pp., Izd. MGU, Moscow, 1937.
Strakhov, N. M., Types of climatic zones during the post-Proterozoic history of the earth and their significance for geology (in Russian), Izv. AN USSR, Ser. Geol., 3, 3–25, 1960.
Strokina, L. A., On the comparison of calculated and observed values of the ocean's radiation balance (in Russian), Tr. GGO, 193, 44–47, 1967.
Suarez, J. M., and I. M. Held, The effect of seasonally varying insolation on the simple albedo-feedback model, Proc. WMO/IAMAR, Symp. on Long-Term Climatic Fluctuations, WMO, 421, 407–414, 1975.
Takahashi, Carbon dioxide cycle in the sea and atmosphere. The Encyclopaedia of Atmospheric Sciences and Astrogeology, Ed. R. W. Fairbridge, Reinhold Publ. Corp., New York-Amsterdam-London, 131–135, 1967.
Teis, R. V., and D. P. Naidin, Paleothermometry and isotope contents of oxygen in organogenic carbonates (in Russian), 255 pp., Nauka, Moscow, 1973.

The Solar Constant and the Earth's Atmosphere, Proc. Workshop, Big Bear Solar Obs., Big Bear City, 1975.
Tyndall, I., On the absorption and radiation of heat by gases and vapours and on the physical connection of radiation absorption and conduction, Phil. Mag., 22, 144, 167–194, 273–285, 1861.
Velichko, A. A., Natural Process in the Pleistocene (in Russian), 256 pp., Nauka, Moscow, 1973.
Vernadsky, V. I., The Biosphere, 146 pp., ONTI, Leningrad, 1926.
Vernekar, A. D., Long-period global variations of incoming solar radiation, Meteorol. Mon., 12, 34, 21 pp., 1972.
Vinogradov, A. P., Introduction to the Geochemistry of the Ocean (in Russian), 213 pp., Nauka, Moscow, 1967.
Vinogradov, A. P., Atmospheric changes under the influence of human activity (in Russian), Geokhim., 1, 3–10, 1972.
Voeikov, A. I., Climates of the World, particularly of Russia (in Russian), St. Petersburg, 1884.
Voeikov, A. I., Climatic fluctuations (in Russian), Meteorol. Vestn., 1, 1–12, 1902.
Wallace, A. R., Origin of human races and the antiquity of man deduced from natural selection, Anthrop. Rev., 2, CLVIII, 1864.
Wallace, A. R., Geological climates and origin of species, Quart. Rev., 126 pp., 1869.
Washington, W. M., On the possible use of global atmospheric models for the study of air and thermal pollution. Man's Impact on the Climate, pp. 265–276, MIT Press, Cambridge, Mass., 1971.
Washington, W. M., Numerical climatic-change experiments; the effect of man's production of thermal energy, J. Appl. Meteorol., 11, 5, 768–772, 1972.
Weather and Climate Modification, Washington, 1966.
Weatherald, R. T., and S. Manabe, The effects of changing the solar constant on the climate of a general circulation model, J. Atmos. Sci., 32, 11, 2044–2059, 1975.
Wexler, H., Radiation balance of the Earth as a factor of climate changes. Climatic change: evidence, causes and effects, Harvard Univ. Press, 1953.
Wexler, H., Variations in insolation, general circulation and climate, Tellus, 8, 480–494, 1956.
White, O. R., Sun. The encyclopaedia of atmospheric sciences and astrogeology, ed. R. W. Fairbridge, Reinhold Publ. Corp., New York-Amsterdam-London, 1967.
Williams, J., R. G. Barry, and W. M. Washington, Simulation of the climate at the last glacial maximum using the NGAR global circulation model. Inst. Arctic and Alpine Res., Occas. Paper, 5, 23 pp, 1973.
Wilson, A. T., Origin of Ice Ages: an ice shelf theory for Pleistocene glaciations, Nature, 201, 477–478, 1964.
Wüst, G., Oberflächensalzgehalt, Verdunstung und Niederschlag auf dem Weltmeere. Länderkundliche Forschung, Festschrift Norbert Krebs, 1936.

Yamamoto, G., and M. Tanaka, Increase of global albedo due to air pollution, J. Atmos. Sci., 29, 8, 1405–1412, 1972.

Yudin, M. I., The influence of forest strips on the turbulent exchange, and their optimal width, in Problems of Hydrometeorological Efficiency of Afforestation for Field Protection (in Russian) 84 pp., Gidrometeoizdat, Leningrad, 1950.

Yudin, M. I., On the possibilities of controlling large-scale atmospheric motions in Contemporary Problems of Climatology, (in Russian), pp. 393–412, Gidrometeoizdat, Leningrad, 1966.

Zeuner, F., The Pleistocene Period, Hutchinson, London, 1959.

Zubenok, L. I., Water Balance of the Continents and Oceans (in Russian), Dokl. AN USSR, 108, 5, 829–832, 1956.

Zuev, V. E., L. S. Ivlev, and K. Ya. Kondrat'ev, New results in studies of atmospheric aerosol (in Russian), Izv. AN USSR, Fis. Atmos. Okeana, 9, 4, 371–386, 1973.